P9-AGP-129

THE BARD ON THE BRAIN

THE BARD ON THE BRAIN

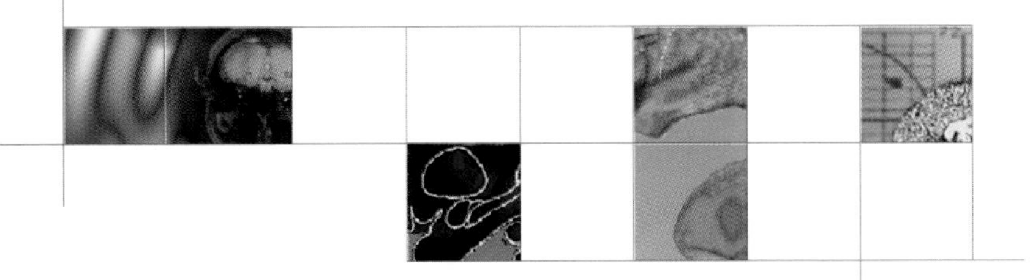

Understanding the Mind Through the Art of Shakespeare and the Science of Brain Imaging

Paul M. Matthews, M.D.
and Jeffrey McQuain, Ph.D.

Foreword by
Diane Ackerman

THE DANA PRESS
NEW YORK • WASHINGTON, D.C.

THE DANA PRESS
745 Fifth Avenue, Suite 900
New York, New York 10151

Editorial offices:
The Dana Center
900 Fifteenth Street NW
Washington, D.C. 20005

Library of Congress Cataloging-in-Publication Data

Matthews, Paul M.
The Bard on the brain : understanding the mind through the art of Shakespeare and the science of brain imaging / Paul M. Matthews and Jeffrey McQuain ; foreword by Diane Ackerman.
p. cm.
Includes bibliographical references.
ISBN 0-9723830-2-6 (alk. paper)
1. Shakespeare, William, 1564-1616—Knowledge—Psychology. 2. Drama—Psychological aspects. 3. Mind and body in literature. 4. Psychology in literature. 5. Emotions in literature. 6. Brain—Imaging. I. McQuain, Jeff, 1955- II. Title.
PR3065 .M38 2003
822.3'3—dc21

2002151166

THE DANA PRESS is the publisher for The Charles A. Dana Foundation

Designed by FTM Design Studio
Printed in Korea

Acknowledgments

This book has been a pleasure to write. It has allowed us to more deeply enjoy the beauty of Shakespeare, while stepping back to look with wonder on the accomplishments of modern brain science. However, neither of us would have dared to take on the challenge of linking these two individually complex topics had Jane Nevins, editor in chief of the Dana Press, not so persuasively outlined the scheme. Her enthusiasm was infectious. It both started the project and sustained it through the difficult periods when we were struggling to develop a way of trying to make the connections exciting yet plausible.

In addition, we wish to thank the research staff of the Centre for Functional Magnetic Resonance Imaging in the University of Oxford for their generous contributions of time and images to the project. We also wish to thank the several other scientists who have contributed images and ideas to the effort. For the wording of the Shakespeare passages, we are indebted to the New Arden edition of the plays, and we also thank William Safire for his guidance and support of the project. Thanks to Shakespeare theatres on both sides of the Atlantic who helped locate wonderful performance photos, and a special nod to the Shakespeare Theatre in Washington, DC for their thoughtful responses to Dana Press's many inquiries. Thanks also to Dana Press director of production, Randy Talley and editor J. Andrew Cocke, art researchers Katherine Hjerpe and Betsy Thompson, and to the designer extraordinaire, Francesca Moghari, all of whose great care has helped make the final book so handsome. Finally, the editorial guidance of Hilary Hinzmann was greatly valued, as was the intelligent copy editing of Anne Nolan, who has done much to transform our often-inelegant prose into something at least readable.

Paul M. Matthews and Jeffrey McQuain

Contents

1. Minds and Brains

2. Seeing, Smelling, Feeling

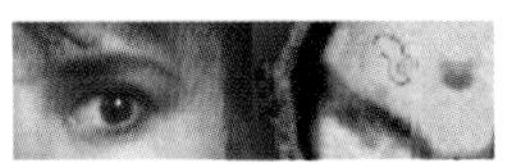

3. Decision and Action

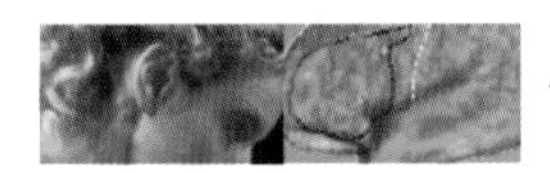

4. Language and Numbers

5. Our Inner World

6. The Seventh Age of Man: Disease, Aging, and Death

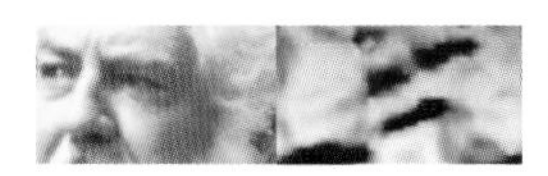

7. Drugs and the Brain

Foreword

On the Bard's Brain

"Beware of Shakespeare!" a man said to me recently. "You can't trust any of the characters. Othello is charming, but too fruity. Pretty Jessica is dependable, but downright common. Jacquenetta is appealing, all right, a real buxom country wench, but completely unstable. Prospero can be subtle, with an interesting spectrum of moods, but just doesn't appear for long enough. Proud Titania, when you come down to it, has too many problems to keep track of. For my money, William Shakespeare is a gaudy giant, but requires altogether too much work."

That did it. "Listen," I said, "I think they're all brilliant creations. Okay, you don't get much of Prospero, but what you do get is rich and unforgettable. Sweet Juliet has the sort of blushing extravagance I would defend to the grave!"

We were two gardeners talking about roses. Wouldn't Shakespeare be surprised to find many of his characters transmogrified in gardens all over the world? He'd probably hate the bad press the David Austin English roses with Shakespearean names are getting at the moment. I'm not sure how he'd feel about all the restaurants, products, and paraphernalia named after him and his works. The admiration of countless readers he'd welcome. After all, a favorite argument in his sonnets is that his beloved will live eternally in the poems, which will be eternally loved. But Shakespeare foods and toiletries? On-line Shakespeare sites, where visitors vote for their favorite play? Entire libraries devoted to the mysteries of his life and the majesty of his works? I believe those would give him pause.

What I think would delight, frighten, and fascinate him at the same time is the simple, universally accepted truth that, when it comes to artistic genius, he stands alone. There's Shakespeare, followed by a very large gap, and then all the other English writers who have ever lived. No one could write like him, not even such renowned stylists as Browning or Nabokov. Something about his brain was different.

Familiar enough to illuminate the human condition in recognizable, entertaining, and profound ways, but different enough to do it in ways and words no one else could achieve. Something about the radar net of his senses was different. Something about his ability to combine seemingly unrelated things in a metaphor's alchemy was different. His ability to juggle many swords of insight at the same time was different. In truth, the people of his era had a very small vocabulary; ours is exponentially larger. But his gift didn't require more words, because words, being human made, can't begin to capture the experience of being alive or the complex predicaments even simple people get into. Words are small shapes in the formless chaos of the world. They're unwieldy, sloppy, even at their most precise. Nothing is simply blue. No one just walks. Words fail us when we need them most. They fall between the crevasses of feelings. If we make them overlap, then we can bridge some of those spaces, and that's traditionally what writers, especially poets, do. A metaphor is hypergolic, like nitroglycerin or table salt. It takes two otherwise harmless things, smacks them together, and creates something explosive. How clever of the brain to find such an enchanting solution.

The thing is, Shakespeare did this more accurately than anyone. It's not just that his senses were unusually keen, though they were, or just that he was a patient and detailed observer of human nature, though he was. He must also have possessed a remarkable general memory, the ability to obsess usefully for long periods of time (*usefully* being the operative word here; he probably could obsess uselessly too), a superb gift for focusing his mind in the midst of commotion, quick access to word and sense memories to use in imagery, and a brain open to novelty and new ideas. His personality must have included these traits, too, because they're essential to all creativity: perseverance, resourcefulness, a willingness to take risks, the urgent need to make exterior his inner universe, the ability to live not only his own life but also the life of his time, a mind of large general knowledge and strength that could be drawn to some particular phenomenon or problem, the capacity to be surprised, passion, the innocent wonder of a child made available to the learned, masterful adult. That all these qualities, and many more, might combine to produce what we refer to as a moment of inspiration is one of the great mysteries and triumphs of mind.

His was a dangerous era in which to be a playwright at court. Elizabeth I and James I attended private performances of his plays, and he had to be tactful about what he said. I'm sure he would have understood the embroidered pillow I once saw in a shop window in Palm Beach that read: "Be careful. The toes you step on today may be connected to the butt you have to kiss tomorrow." But spontaneity bound by restraint is the stock and trade of creativity. It's one of the brain's favorite ways of creating art. Thus paintings have frames, symphonies and verse forms have strict rules. The restraints aren't always political. Sometimes the medium has built-in limits—oil paint mixes and dries in characteristic ways. As technology changes, limits sometimes change, as when oil paint became portable. Some restraints are ordained by society.

For example, in Shakespeare's day, arranged marriages were still the norm, but many people started objecting to them. His plays are filled with commotion over whom to marry, the right to choose a mate, and complaints by characters who'd prefer a love match. The best known of them, *Romeo and Juliet*, was a classic told in many cultures and genres when Shakespeare decided to tell the story yet again, doing what Leonard Bernstein did with *West Side Story*—adapting a well-known, shopworn tale to contemporary dress, locale, and issues.

He created a younger Juliet than in other versions, allowing a beautiful thirteen-year-old Veronese girl to encounter the embodiment of her robust sensuality: a boy who is passion incarnate, someone in love with love. "Love is a smoke made with the fume of sighs" (1.1.190), he tells his friend Benvolio. Romeo is a bolt of lightning looking for a place to strike. When he meets Juliet the play's thunderstorm of emotions begins. All versions of the story hinge on the rivalry between two noble houses, and the forbidden love of their children. In Shakespeare's, chance, destiny, and good playwriting ordain

Romeo and Juliet. Derek Smith as Romeo and Laura Hicks as Juliet.

that they shall meet and become "star-crossed lovers" with a sad, luminous fate. The use of lightning and gunpowder images throughout the play keeps reminding us how combustible the situation is, how incandescent their love, and how life itself burns like a brief spark. Full of tenderness and yearning, their moonlight balcony scene contains some of the most beautiful phrasing ever written, as they sigh for love under the moon and stars, most alive in a world of flitter and shadow.

A director friend told me recently that while watching news coverage of war-ravaged Bosnia one evening, he began thinking differently about *Romeo and Juliet*. Suddenly he realized that the play isn't really about star-crossed lovers. It's about what happens to children in a culture of violence, and that's how he directed it at the Shakespeare Festival in Kansas City. Because Shakespeare excelled at constructing a house of cards, whose walls are intersecting planes of meaning that support one another, both interpretations ring true.

Another angle on Shakespeare's brain is that he wasn't good at inventing plots. He elaborated them cleverly once he had them, but for the most part he borrowed plots from historical

Romeo and Juliet. Derek Smith as Romeo and Edward Gero as Tybalt.

sources. As I understand, sadly, plotting requires a special cast of mind. Give me a ready-made plot and I'll have fun elaborating it. Ask me to make phrases until the cows come home, and I'm happy. Ask me to describe a gesture or set a scene or develop an idea or explore someone's psychology, and I'm happy. But ask me where the person crossing the room just came from, and I have no idea. It's a mechanically different process. To do it easily takes a brain tilted in a slightly different way. An example of this is the potboiler, a one-dimensional book that has nothing to do with literature, although its plotsmanship may be brilliant.

Style flows from the sediment of one's personality. It's hard to know what Shakespeare really believed, because his different characters contradict one another. I've written dramatic monologues and a verse play, and there are reasons poets find them compelling to write. Here are a scant few: Sometimes one writes passionate soliloquies about feelings one is only testing out, to probe them, to see what they would feel like, painfully, enthusiastically—but not irrevocably. For reasons I won't even try to explain, writers often feel the need to be private in public, and taking on the mask of a dramatic monologue does that well, as a form of ventriloquism. One can splurge on emotions like anger or unrequited love, reveal real and imaginary crimes. One can say politically dangerous things. Dramatic monologues give one license to be licentious, yet one can disown the feelings as mere inventions, the ravings of a buffoon created for stage effect.

Style is also a penetration of the world, an adventuresome safari through daily life. The drafter of "O that this too too sullied flesh would melt" has probably witnessed the slow putrefaction of pork as well as ice melting on the local pond at Warwickshire. The line is a blurt, an educated man's yearning to be nobody (rather than a young prince obsessed with his royal father, whose image he carries about with him). The energy of that statement comes from a Shakespeare who has opened his own soul to the possibility of evaporation.

A self is a frightening thing to waste. It's the lens through which one's whole life is viewed, and few people are willing to part with it, in death, or even imaginatively in art. Shakespeare was a master of empathy and surrender, who gave himself to the human dramas he saw daily, figuring out how he would feel if doing or saying certain things. He became a crowd. All alone, he could have been arrested for unlawful assembly. I don't imagine this is something one turns on and off very easily, or suddenly decides to try; he probably imagined himself as others often, secretly, from childhood on. Why he would have found escape from self necessary, or fun, is not for me to speculate. Most people won't be pried loose from the single word *I,* which is, in the end, our only possession. He savored human phenomena, humans as intriguing sensory suits that he borrowed now and then. With equal measures of grandiosity and humanity, Shakespeare would slip out of his self and into another's, then quiz his senses and cast the feelings and sensory information into his best language. I wonder how many Autolycuses he knew personally before he came up with the definitive phrase "a snapper-up of unconsidered trifles."

Shakespeare had the courage of his diffidence. He soaked things up and brooded on them before allowing them to fill the mouths of Macbeth, Portia, or Prospero. Of course, to be so giving in your response to life, you have to be mighty sure of yourself, aware that you will always be able to whip out a unique phrase that will displace all the efforts of all the other authors who don't achieve anything like sixty-odd pages in *Bartlett's Familiar Quotations*. You also have to be willing to experience extremes of emotion, no matter where they lead, even as far and low, say, as that moment in *King Lear* when with hammer blows he decrees the extinction of Cordelia: "Never, never, never, never, never." Five *nevers* is a lot. The words function as individual guillotines, and there's no question about the anguish felt in the writing of them. It is the numb bleat of an author who diagnoses individual doom.

At the end, Hamlet beseeches Horatio to report him aright in the world, to paint an accurate picture; he knows no celestial scribe is going to do the job for him, or the nonexistent media of the day. Word of mouth is all, and as Shakespeare foretold in his sonnets, it has served him well over the centuries. It's not enough to say that a certain play of Shakespeare's is full of quotations, or even, as George Bernard Shaw said of *Henry V*, that it's a national anthem in five acts. The mind at work in all the plays has a unique idiom. Something was wonderfully different about his brain.

The texture of his imagination favored certain rhythms and patterns. "The bank and shoal of time" is one of his treasured rhythms: the one and two of three. When you think about it, that's a convoluted way to build a metaphor. In "the slings and arrows of outrageous fortune," for instance, items owned by an unstable human archer (slings, arrows) imply the calamitous actions they're capable of. And then the phrase says: "Okay, now swap fortune for archer." A listener hears the first part, allows the items to conjure up the man, pictures the archer, then is surprised by the powerful appearance of something abstract—fortune—which has now acquired familiar and scary human characteristics. All that in a phrase heard fleetingly, and yet our brains follow the trajectory of the imagery without a sweat, and we're powerfully moved by it. Shakespeare has managed to

Hamlet. Kevin Kline as Hamlet.

endow a superstitious force with a ferocity we have seen, or perhaps even possessed.

There are some rhythms used successfully by other poets—Browning, say, or Swinburne—that Shakespeare didn't seem to care for at all. Iambic may be the natural rhythm of someone walking, but how many people can explore a mood in blank verse? His thoughts had a unique cadence. Some of his characters talk to themselves in sonnets. It's appealing to use the straitjacket of a verse form to organize intense emotions, and I'm sure he worked hard at word choice, as all professional writers do, especially ones with appalling deadlines. But, nonetheless, his mind seemed to entrance itself with verse. I wonder if his favorite rhythms didn't work as a reliable sort of incantation, which focused the mind, disrupted the mundane hodgepodge of thoughts that one needs to buy bread or shoot the bull with pals, and spirited him away to another mental kingdom.

Did Shakespeare know how different he was? Probably so. Even in idle chitchat, people report on what they feel and see, revealing how the world touches them and the realm of their sensibility. He would have known how alien he was. How human, in a hundred familiar ways, but also how different. It would have been both his privilege and burden to be extraordinary. More of everything. More hearing, seeing, feeling, smelling, more imagining, and perhaps more hurting, thrilling, angering. What was the texture of his imagination? Would mapping his brain have told us? If he were alive today, could scientists begin to explore the mountains of his mind? I wonder.

Recently, longing yet again to know who Shakespeare the man really was, I phoned the Shakespeare specialist at an Ivy League university and asked her advice on which biography to trust.

"Oh, I'm not interested in the man or his work," she said.

"I see," I said, pausing until the shock wore off enough that I could continue speaking. "Well, if you're not concerned with the man or his work, then what aspect of Shakespeare are you interested in?"

"The political response to the canon," she said decisively.

"And that's what you teach?"

"Of course."

"Not the man or his work?" I asked this in as neutral a tone as I could, since I sincerely wanted to know her answer.

"No!" she said impatiently, and her voice had a condescending edge.

Clearly, if I wanted any information from her, I would need to apologize for my aberration. "Well, perhaps you'd humor me, you know how writers are—I'm enchanted by the work and would like to learn more about the man." She suggested a biography that relied heavily on legal documents from Shakespeare's time. It was a useful suggestion. But when we hung up, I felt enormous grief, pity for the students who would find Shakespeare's work reduced to the tidy annals of political theory, and a profound, visceral sense of desecration. Not because Shakespeare was a god. If anything, he risked being more human than most. Because he was a natural wonder.

Diane Ackerman

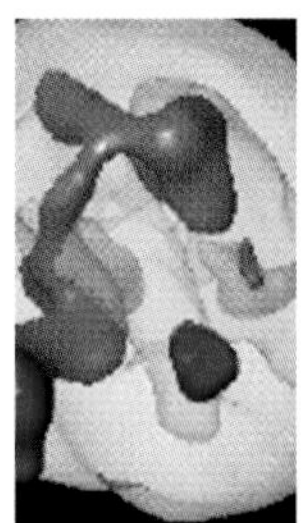

Introduction

For the people of England, the Elizabethan period was a time of enormous expansion in many spheres of life. The broadening of intellectual horizons throughout Europe established the foundations for the period of Enlightenment in the eighteenth century. It was a time of great economic and social change, during which England became a dominant European power. A key to this power was control of the seaways, by which explorations of the outside world proceeded ever more rapidly. And as new lands were being discovered, the nature of the European world was being redefined and the sense of human potential enlarged.

William Shakespeare both led and reflected his age. He developed the English language to an extent that no single writer has since. He mined the language of the rich and poor, rulers and the ruled, to develop more precise ways of expressing his thoughts and feelings. In doing so, he explored the inner world of man in a way that paralleled the journeys of the seafarers whose tales filled the taverns of England's ports.

Shakespeare was a keen observer of human nature. In his plays he worked to define the minds of his characters so as to explain why they act as they do. His stories are timeless not because of the originality of the plots (which generally were borrowed from earlier writers) but because they thoughtfully explore questions about the human condition so fundamental that they continue to absorb us today.

Recall, for example, how in *As You Like It,* Shakespeare distills the poignant brevity of human life in the twenty-eight lines of poetry in which the character Jaques recounts the seven ages of man. Shakespeare's 154 sonnets capture the experience of love from the dizzying perspectives of both its heights and its depths. And Hamlet's soliloquy beginning "To be, or not to be . . . " (*Hamlet,* 3.1) explores the darkness of depression and suicide. From the vivid accounts of the birth of Queen Elizabeth I in *Henry VIII* to the death of the elderly King Lear, Shakespeare's masterful descriptions of human growth and aging remain unequaled.

What we appreciate as Shakespeare's genius derives from his keen insight into the human mind and from his obvious excitement in using this insight to experiment in drama. While his experiments were not designed and executed as are those of modern brain scientists, the underlying goals had intriguing similarities. His laboratory was the theater, where he tested his words and refined them until they communicated powerfully and accurately. Like a modern brain scientist, he was testing hypotheses concerning the ways in which the human mind works. By using—and at the same time working to define—this complexity in his poetry and plays, he achieved his great art. In creating his enduring theater, Shakespeare also defines for us the uniqueness and wonder of the human mind.

As the twenty-first century begins, we also are living in a time of great change and of startling advances in human knowledge. Like that of the Elizabethans, our world is broadening, although perhaps we have entered the twenty-first century looking more to the world inside than outside of ourselves. One of the crowning achievements of the last century was the development of a substantial understanding of the ways in which our thoughts and feelings arise out of activity in the physical substance of our brains. In much the same way that the Elizabethans' discovery of the world was made possible by advances in the technology for seafaring, so recent scientific advances have moved forward with development of new technologies that allow us to observe the brain at work.

The modern era of brain science began in the late nineteenth century with improved microscopes and tissue-staining techniques. These allowed a detailed characterization of the arrangement and nature of the cells from which the brain is built. New tools have allowed the focus to move toward understanding the basis of thoughts themselves—the mind that is generated from the brain. In attempting to understand the mind, brain scientists finally have the means to address questions that Shakespeare so eloquently put forward four centuries ago. What is imagination? How do our memories develop? What is it about a stirring speech that motivates us? Why is it hard to learn a foreign language?

Shakespeare was concerned not only with the normal brain; he also observed keenly brain diseases that challenge us now. Shakespeare had a particular interest in the mind gone wrong. He probed aspects of madness with Hamlet, Ophelia, and Lady Macbeth, to name but a few instances. In *King Lear,* Shakespeare addressed the problems of aging and dementia. In *Julius Caesar*, Casca recounted the way the crowds wondered at Caesar's having an epileptic seizure in the market square.

Shakespeare noted the way drugs change the mind and behavior. In several places, he asked what it is that alcohol and other drugs do and why they are so prone to abuse. There is no more concise description of the sad effects of alcohol abuse than that of Cassio in *Othello* (2.3): "O God, that men should put an enemy in their mouths, to steal away their brains! that we should with joy, pleasance, revel and applause, transform ourselves into beasts!" Shakespeare's cry for effective treatments for disorders of the brain is echoed by the families and doctors of patients who remain ill despite the best available treatments: "Canst thou not minister to a mind diseas'd?" Macbeth asks the doctor attending his disturbed wife, "Cleanse the stuff'd bosom of that perilous stuff/Which weighs upon the heart?" (*Macbeth*, 5.3).

In a very modern way, Shakespeare appreciated that there is sometimes only a fine line between sanity and madness. He sought the roots of madness in the makeup of the individual. Like Freud, Shakespeare explained madness in psychodynamic terms. Possible causes of Hamlet's supposed madness, for example, are sought by Claudius the king and the ill-fated Polonius as they eavesdrop on Hamlet's meeting with Ophelia.

Part of the current excitement in brain science arises because scientists now have tools that can define the brain mechanisms responsible for the thoughts, emotions, and disorders that Shakespeare so wonderfully described. In recent years, noninvasive recordings of brain electrical activity using electroencephalography (EEG), or recordings of the associated tiny magnetic fields using magnetoencephalography (MEG), have allowed the action of nerve cells in the brain (neurons) to be defined (Fig. 1). The chemical changes in the neurons that make their electrical activity possible can be mapped using techniques such as positron-emission tomography (PET; Fig. 2) or single-photon-emission computed tomography (SPECT). These

Figure 1. *These bright images are representations of the extremely low voltage electrical activity recorded on the surface of the head using electroencephalography (EEG) during a thinking task. The task involved reading individual words for meaning. Each brain image shows the electrical activity at a precisely defined time after a word was presented. The changing patterns (shown by the shifts in position of bright areas) identify the sequence in which different regions of the brain become active during the reading of a word.*

methods can also map changes in blood flow that occur during cognitive processing as they track the distribution of small quantities of radioactive chemicals injected into the blood for this purpose. In some cases, the chemical signaling between neurons can also be measured using PET.

One of the most recent techniques to be introduced into the brain sciences is functional magnetic resonance imaging (fMRI;

Fig. 3). It relies on detecting tiny increases in the amount of fresh blood flow when the brain becomes active for perception, movements, or thought. This technique has the particular advantage of providing a clear picture of the brain without the use of radioactive chemicals. It is also significantly quicker than PET—the fMRI method can map the blood flow responses that accompany brain activation within seconds.

We hope that this book will help you appreciate how Shakespeare's interest in the human mind has parallels in the

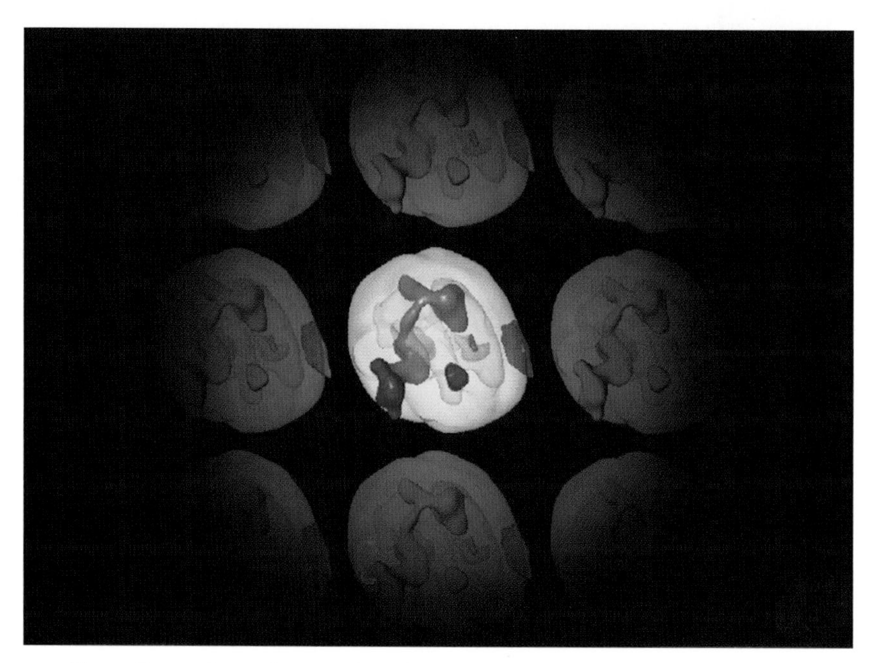

Figure 2. *Bright red is used to define areas of the brain that become active after application of painful heat in this positron-emission tomography (PET) image. To generate this type of PET brain image, subjects were given a very small amount of a radioactive material to trace blood flow in the brain, first when they were lying comfortably, and then again when a painfully hot stimulus was applied. Differences in the two "maps" of blood flow obtained before and after the heat were then used to generate this image, which shows areas of the brain that become active with pain.*

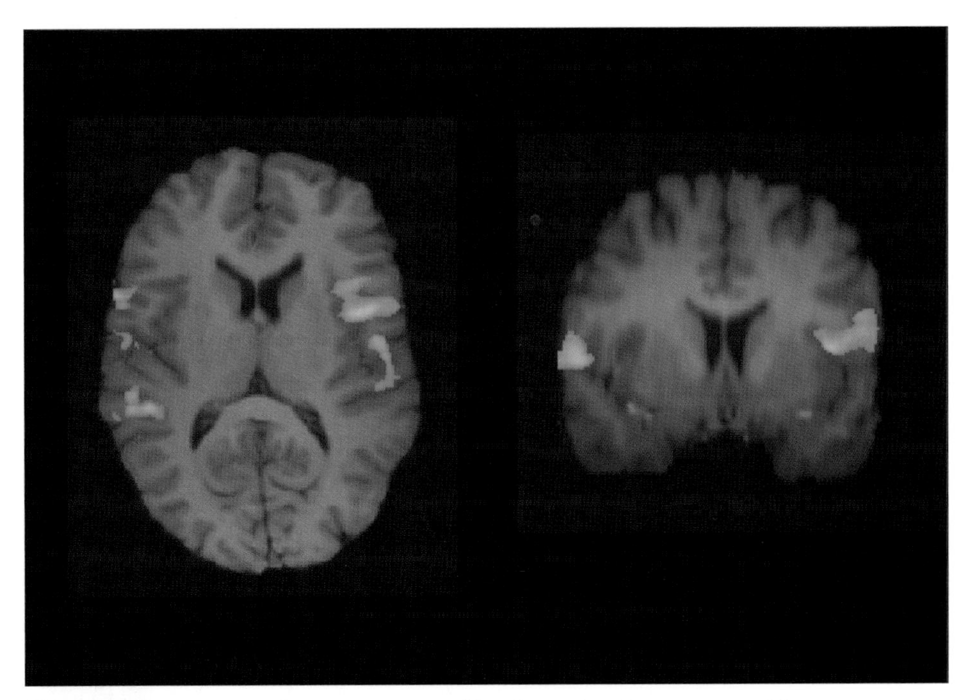

Figure 3. *This picture shows patterns of brain activation measured by functional magnetic resonance imaging (fMRI), which measures the small changes in MRI signal intensity that accompany increases in blood flow in specific areas of the brain that become active during a thinking task. In this case, the subject was gently touched on the foot, either when asked to pay close attention to the touch, or when distracted by a counting task. Differences in brain activity were found in many parts of the network involved in sensation, including the primary sensory area. This demonstrates that attention modifies the response to stimuli at the most fundamental level of cortical sensory processing.*

work of brain scientists today. Opening with Miranda's marveling in recognition of the glorious range of humanity, the first chapter provides an overview of the "brave new world" of brain science. The next chapters explore ways in which the brain works to make us aware of the world around us and allows us to act in it. Following this, we address a theme that echoes from the words of Shakespeare throughout the entire book: the awesome power of language. We then turn to discussions of imagery, memory, and emotions. Shakespeare was particularly fascinated by the inner world of the individual mind and the way the imagination can build what Prospero in *The Tempest* calls "cloud-capp'd towers, the gorgeous palaces,/The solemn temples, the great globe itself" (4.1). We close with problems of the disordered brain and the ways in which it is altered by drugs. Like us, Shakespeare watched the tragedy of brain disease with both fascination and sadness.

We stand poised at the dawn of a new Age of Enlightenment that is in part being led by our growing ability to observe brain processes directly. In celebration of this, almost 400 years after Shakespeare's death, science and art—the art of Shakespeare and the science of brain imaging—come together in this book.

A Map of the Brain

These are the important areas of the brain that modern brain imaging now allows us to see as they are working. These regions become active to enable thought, emotions, our senses, and the workings of the body, from the beating of the heart to the movement of a finger.

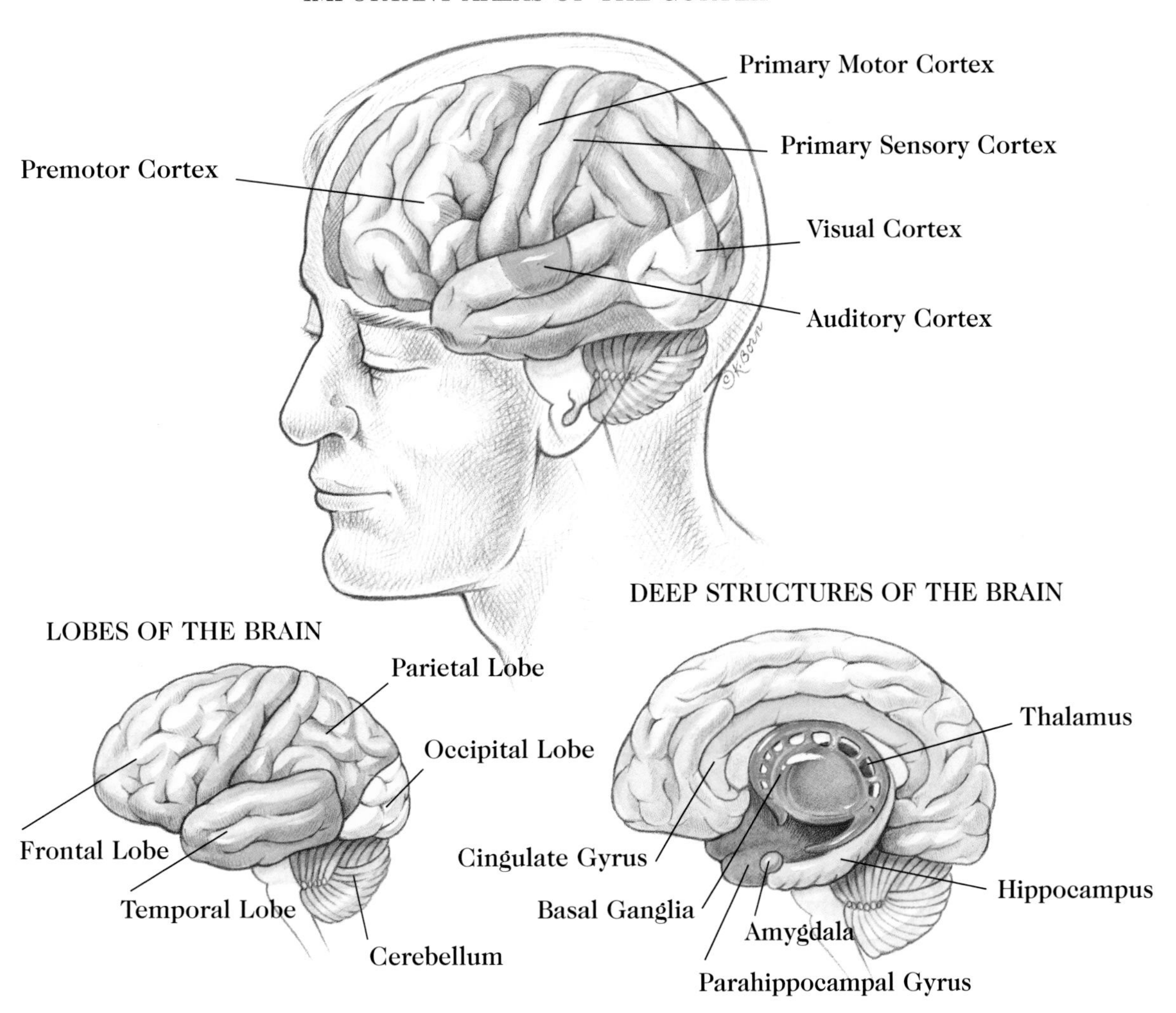

1.  Minds and Brains

THE WONDER OF THE HUMAN BRAIN

Considered by many to be Shakespeare's final play, The Tempest *offers some of the playwright's most enchanting poetry. From a bewitched island, the magician Prospero causes a storm to shipwreck King Alonso of Naples, Alonso's brother Sebastian, and Prospero's brother, who has usurped Prospero's title as Duke of Milan. Prospero ensures that all on board are rescued, but separates Alonso's son, the handsome Ferdinand, from the others in order to bring him alone to his beloved daughter, Miranda. Before this meeting Miranda has known no human company other than Prospero and the beastly Caliban since she was a young child. Late in the play, Alonso sees Ferdinand and Miranda and rejoices at his son's preservation from the shipwreck. For her part, Miranda is overjoyed by the wonder that there are others like her, and in Ferdinand and his friends she finds handsome ones as well!*

The Tempest. Ana Reeder as Miranda and Ted van Griethuysen as Prospero.

Alonso: If this prove
A vision of the island, one dear son
Shall I twice lose . . . .

Miranda: O, wonder!
How many goodly creatures are there here!
How beauteous mankind is! O brave new world,
That has such people in 't!

Prospero: 'Tis new to thee.

The Tempest, 5.1

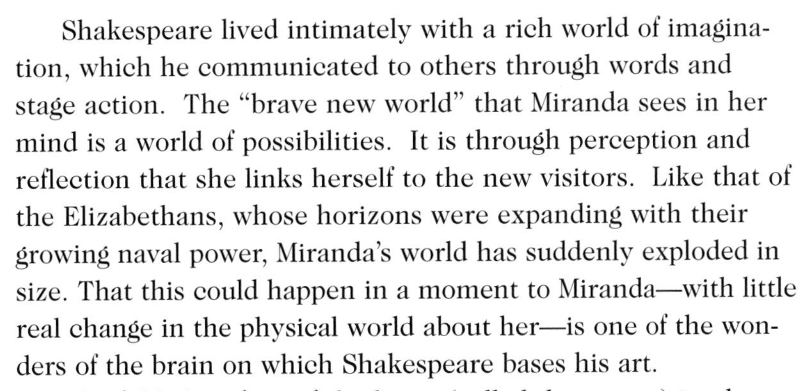

Shakespeare lived intimately with a rich world of imagination, which he communicated to others through words and stage action. The "brave new world" that Miranda sees in her mind is a world of possibilities. It is through perception and reflection that she links herself to the new visitors. Like that of the Elizabethans, whose horizons were expanding with their growing naval power, Miranda's world has suddenly exploded in size. That this could happen in a moment to Miranda—with little real change in the physical world about her—is one of the wonders of the brain on which Shakespeare bases his art.

How beauteous mankind is! O brave new world, That has such people in 't!

The folded surface of the brain (called the cortex) is where neurons are found. Here lie the major mechanisms that allow learning and thus both intellectual and moral growth. Shakespeare explored the theme of development in *The Tempest* in several ways. Most obviously, the young Miranda is being observed during her early emotional maturation into womanhood. We feel a special tenderness toward Miranda because her lack of experience of the world makes her still a child in many ways.

Here Shakespeare appeals to our own direct experiences of the slow maturation of the human mind and brain. The human brain is much slower to develop than that of most animals. It increases more than fourfold in size from birth to the adult average volume of 1,400 cubic centimeters (about 85 cubic inches), about as big as a medium-size cantaloupe. The cortex continues to enlarge after birth not because more cells are added but because branching arms on the nerve cells expand to accommodate the 10,000 or so connections that eventually reach each neuron in the mature brain. In addition to changes in size, a key anatomical feature of this maturation, which can be followed using brain imaging methods like MRI, is the development of a special coating (myelin) around the long connections (axons) between neurons. This coating allows the very rapid conduction of electrical signals between neurons that is necessary for quickness in thought and action. You can appreciate that in an organ with 28,000,000,000 cells and more than 10,000,000,000,000 connections (thousands of times greater than the entire population of this planet!), it is important that the "wiring" be efficient.

However, not only Miranda's development is chronicled in the play. *The Tempest* also tells us how her mysterious father, Prospero, has grown in scholarship to become a great wizard. Intellectually vigorous himself, Shakespeare must have reveled in how much a man could continue to mature emotionally and grow intellectually even into old age. A remarkable aspect of the human brain is its tremendous plasticity, which allows learning and adaptation to occur throughout life. But unlike the development that occurs early in life, these changes in the adult brain occur not so much because of changes in its overall structure but because of changes in the way it functions. The evidence of his plays suggests that Shakespeare closely studied the process by which human beings learn and change their behavior.

Modern technology allows us to directly study changes in the way the brain does its work. Figure 4 illustrates how the patterns of neuronal activity in the brain associated with thought can be followed using the new functional imaging techniques. Here active areas detected by PET are represented as orange spheres, and fMRI activations are shown as green spheres. The spheres are arranged to reflect their relative locations in the brain.

Using MRI, images of the structure of the brain have been created at the sides. In this case, the images record activations in areas of the brain associated with movement. Planning centers in the middle of the brain (in a region known as the supplementary motor cortex) are engaged first, to initiate the direct commands to move that come from the primary motor cortex on the side of the brain. The accuracy of the movement is

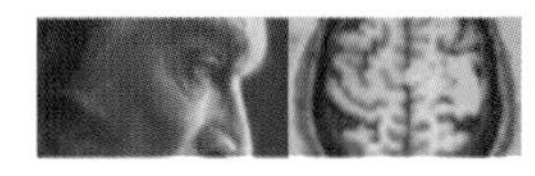

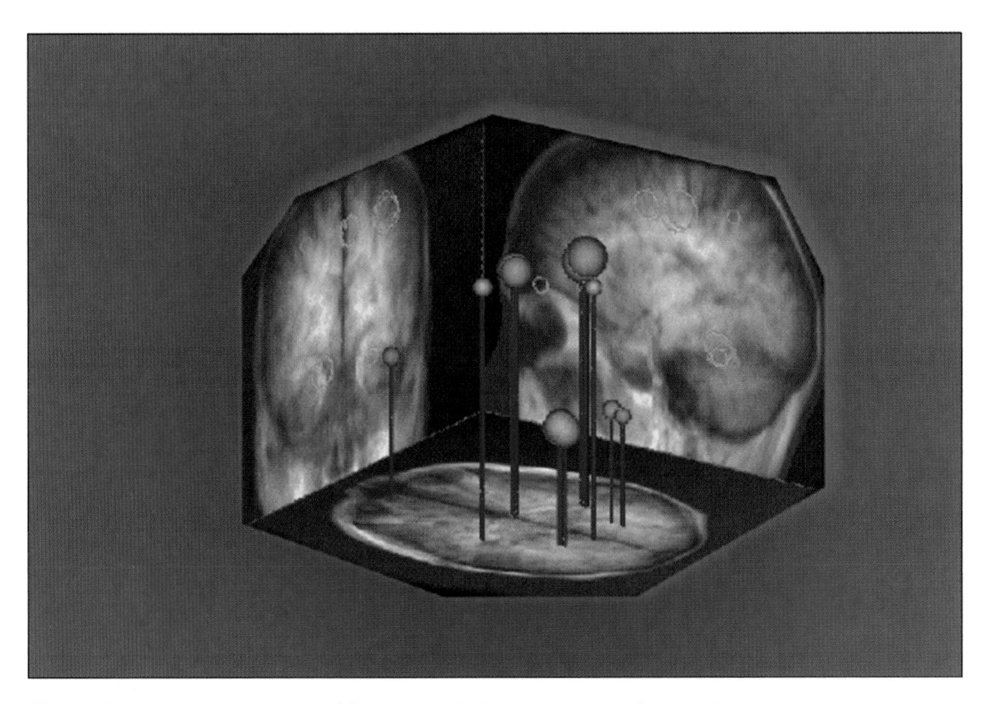

Figure 4. *A representation of functional changes in the brain that occur when a normal subject moves fingers of the left hand. Conventional magnetic resonance imaging (MRI) showing the structure of the brain are shown at the sides (gray). Brain activity during this finger movement task is represented in the central space as balls of either orange (representing results from PET scanning during the task) or green (representing results from fMRI scanning during the task). Deciding how best to represent functional imaging data continues to be a challenge for brain scientists. While this novel representation for functional imaging changes has not become popular, it does provide an intriguing way of visualizing brain activity.*

enhanced by feedback from activity in the cerebellum (a smaller "hindbrain" in the lower back of the skull), on the lower edge of the picture.

This is only one part of a description of relatively elementary processes in a single network of brain activity. This type of basic information must be integrated with a great deal of other data to even begin to address problems of cognitive neuroscience of the sort that consideration of Shakespeare raises. We now must define the much more complex interactions between different processing networks. Such neuroscientific "elegance" still looks rather clumsy when juxtaposed with the poetry of Shakespeare—the Bard's way of describing the psychological dynamics of will and action. Even so, as you read this book, we hope you will appreciate the wonder of the "brave new world" being unveiled by modern science.

SEEING THE MAN THROUGH HIS BRAIN

Grasping the skull of a long-dead friend, Hamlet speaks what is perhaps the most misquoted line ever penned by Shakespeare: "Alas, poor Yorick. I knew him, Horatio" (not "I knew him well," as is so popularly believed). The Prince of Denmark's words come during a brief comic scene near the end of the play, when Hamlet meets a garrulous gravedigger. Hamlet does not yet know that his beloved Ophelia has drowned herself or that her grave is being prepared. When the gravedigger produces the skull of Yorick, once a jester to Hamlet's father, Hamlet fondly recalls the jester from his youth, now perhaps two decades past. Presaging the tragic news of Ophelia that is to follow, the skull reminds Hamlet that all of us must eventually die. Hamlet sees his former friend through his skull—a solid shell that can remain even after the brain is long decayed.

Hamlet: Alas, poor Yorick. I knew him, Horatio, a fellow of infinite jest, of most excellent fancy. He hath bore me on his back a thousand times, and now—how abhorred in my imagination it is. My gorge rises at it. Here hung those lips that I have kissed I know not how oft. Where be your gibes now, your gambols, your songs, your flashes of merriment, that were wont to set the table on a roar? Not one now to mock your own grinning? Quite chop-fallen? Now get you to my lady's chamber and tell her, let her paint an inch thick, to this favour she must come. Make her laugh at that.

Hamlet, 5.1

Hamlet. Kevin Kline as Hamlet.

Hamlet's recollection of Yorick from the jester's skull is a reminder of the very special way we view this most human part of our skeleton. Shakespeare uses a skull as a stage prop because it powerfully inspires awe or even fear in evoking an image of the living person in the context of death (Fig. 5). Our images of early man are reconstructed in large part from the shapes of their skulls, for example. The internal surfaces of such ancient skulls can also be used to reconstruct a picture of the complexly folded surfaces of the brains they once contained. In fact, detailed studies of skulls can actually reveal specific individual characteristics of the brain.

Alas, poor Yorick. I knew him, Horatio, a fellow of infinite jest, of most excellent fancy.

Yorick's "most excellent fancy" must, of course, have arisen from the physical substance of his brain. A recurring theme in brain science (which also may have contributed to Shakespeare's beliefs concerning the brain) has been that the structure of the brain reflects its function in a rather direct way. People have even thought that the shape of the brain (which would be reflected in the shape of the inner surface of Yorick's skull) reflects an individual's personality, intellectual capacity, and moral character. This idea gained its greatest currency early in the nineteenth century, when the German physician Franz Josef Gall and his students attempted to systematize a science of *phrenology*. Phrenologists believed that character traits and abilities were highly localized in the brain. The extent to which an individual manifested a trait or skill was thought to increase with the size of the associated brain region. Thus, they argued (although without compelling evidence) that aspects of a person's character (traits such as "cautiousness" and "mirthfulness") could be inferred simply from the pattern of bumps on the head.

We do not now believe that character traits or brain functions can be localized in such a simple way. Instead, it is felt that the brain processes information by means of complex networks of interactions between widely distributed regions of the brain—rather than solely in specific areas that might be reflected in local skull shape. Yet it is also clear that over the course of its development, the brain becomes regionally specialized for different functions. Perhaps one of the clearest examples of this is the way that regions specialized for processing language are highly lateralized to the left side of the brain in most people.

We do not know whether Hamlet visited Yorick's grave on a bright summer's afternoon, in the fog of early morning, or in a

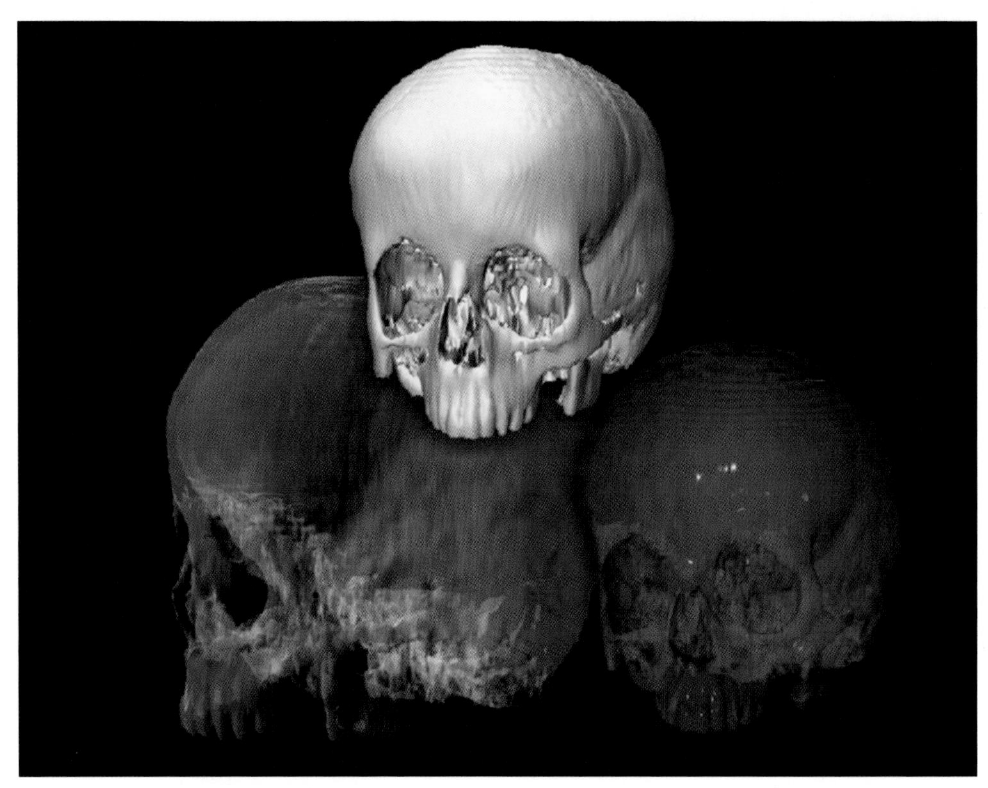

Figure 5. *Computerized tomography (CT) scans use X rays to generate images. These provide an excellent outline of the bony skull. Conventionally, they are shown to doctors as "slices" arranged in order from low in the head to its top. However, for some applications it is ideal to reconstruct the external shape of the skull in fine detail. This can be particularly helpful in reconstructive surgery, for example. These images were generated from CT scans using a special computer reconstruction technique. They clearly show the outlines of the human skull in a living person.*

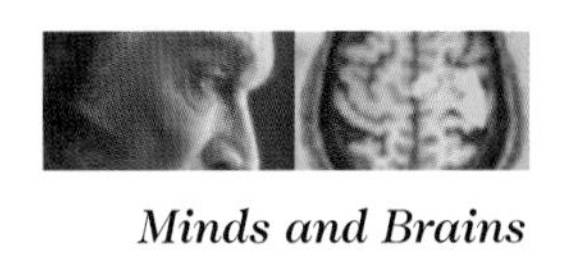

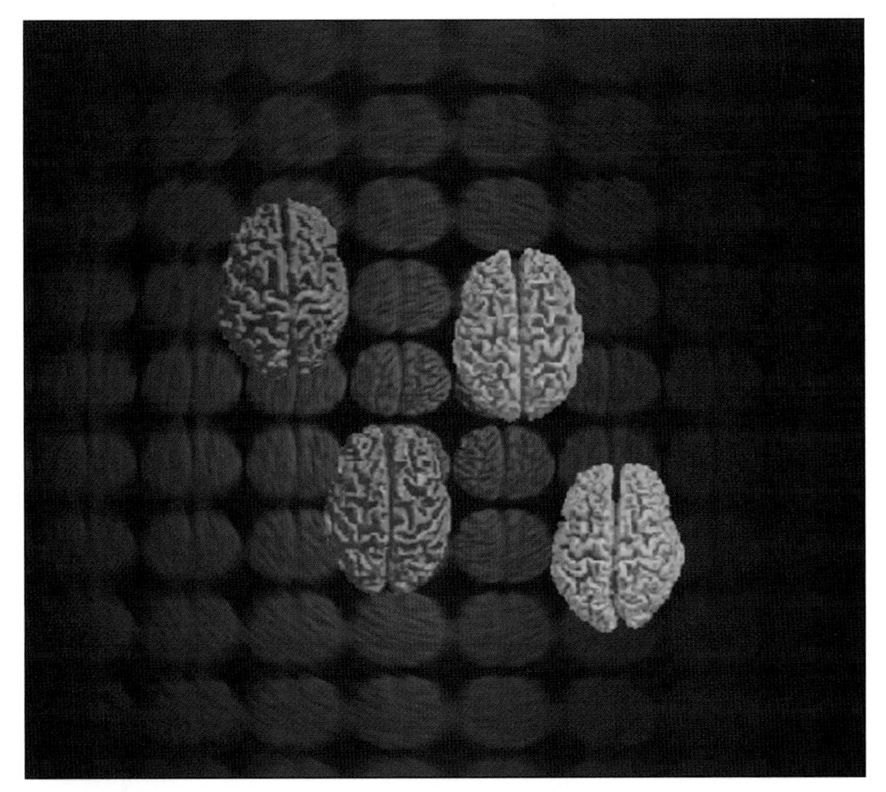

Figure 6. *Individual brains show common major structural features but considerable variation in finer details. Overall shapes may differ in subtle ways, and the number, precise position, and depth of folds in the brain vary among different people. The surface anatomy of the brain therefore carries unique information, just like the fingerprint. Here, MRI scans were reconstructed to make three-dimensional images of four brains, illustrating the differences between the details of their surface structures.*

melancholy gravedigger's dusk, but if he was able to examine the inner surface of the skull closely, he might have seen some marks arising from the general shape of the folds, or gyri, in the underlying brain. The remarkable thing in Hamlet's musings is the way in which he so specifically appreciates this particular skull as having belonged to Yorick. In fact, the patterns of the folds on the surface of the brain are very specific for each individual. While there are general similarities between all people, the precise position, shape, and size of the individual folds seen on the surface of a brain are as unique as a fingerprint.

The second image here shows the surfaces of the brains of four elderly people as reconstructed from their brain scans (Fig. 6). They are representative of any small sample of human brains. At first glance they look very much alike, but on closer inspection substantial differences become apparent. These brains differ in the shapes of the division between the two halves of the brain (the interhemispheric fissure), in the pattern of the grooves in the surface (the sulci), and in the thickness of the individual gyri. An important goal of current brain imaging research is to define these differences precisely and to understand their possible significance in determining individual behavior.

WE SHARE A COMMON HUMANITY

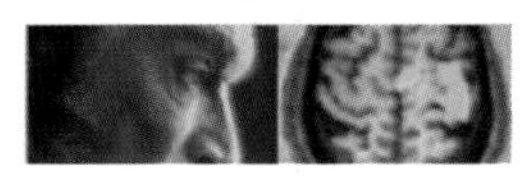

The Merchant of Venice *is perhaps the most controversial of Shakespeare's comedies. It describes a deadly struggle between a Christian merchant, Antonio, and a rich Jew, Shylock. To help his friend Bassanio court the rich heiress Portia, Antonio borrows money from Shylock, agreeing to forfeit "a pound of flesh" if he cannot repay the debt. Unfortunately for Antonio, the profit from his trading ships fails to arrive on time and Shylock, bitter from years of mistreatment by the Christians of Venice, seeks to exact his bloody due. Antonio's friends Solanio and Salerio confront Shylock, and the angry Jew explains how anti-Semitism has made him hungry for revenge. His speech offers Shakespeare's most eloquent argument for the need to recognize the humanity we all share. As this scene opens, Shylock is explaining why he wants to claim the pound of flesh.*

Shylock: To bait fish withal,—if it will feed nothing else,
it will feed my revenge; he hath disgrac'd me, and
hind'red me half a million, laugh'd at my losses,
mock'd at my gains, scorned my nation, thwarted my
bargains, cooled my friends, heated mine
enemies,—and what's his reason? I am a Jew. Hath
not a Jew eyes? hath not a Jew hands, organs,
dimensions, senses, affections, passions? fed with
the same food, hurt with the same weapons, subject
to the same diseases, healed by the same means,
warmed and cooled by the same winter and summer as
a Christian is?—if you prick us do we not bleed?
if you tickle us do we not laugh? if you poison
us do we not die? and if you wrong us shall we not
revenge?—if we are like you in the rest, we will
resemble you in that. If a Jew wrong a Christian,

The Merchant of Venice. Geraldine James as Portia and Dustin Hoffman as Shylock.

> what is his humility? revenge! If a Christian wrong a Jew, what should his sufferance be by Christian example?—why revenge! The villainy you teach me I will execute, and it shall go hard but I will better the instruction.
>
> *The Merchant of Venice*, 3.1

Writing in a time of commonly accepted and long-standing discrimination against Jews, Shakespeare gives Shylock a moving eloquence for his complaints against this injustice. Shylock adds to the case for the common humanity of Christians and Jews by describing their shared physical features and shared features of mind: "hands, organs, dimensions, senses, affections, passions."

I am a Jew. Hath not a Jew eyes? hath not a Jew hands, organs, dimensions, senses, affections, passions?

As noted in the previous section, today's brain scientists are becoming increasingly interested in addressing issues such as the basis of individual behavior and personality. Because it is clear that the complexities of a person's character do not map onto specific brain structures in a simple way, one strategy is to work to identify key common features in the structures of individual brains with more general aspects of individual potential, such as memory capacity or finger dexterity.

It is remarkable that despite the great differences in size, shape, skin color, and other physical features of the bodies of different men and women around the world, the size of the adult brain is remarkably constant. The general shape of the brain is also similar across all human populations, with two relatively symmetrical hemispheres around the central, fluid-filled cavities called ventricles. Other major structural features, such as the deep fold in the middle of each half of the brain (or hemisphere), known as the central sulcus, and the division between the two hemispheres of the brain (the interhemispheric fissure), are also very much the same in all brains. Even the gyri, whose patterns in each individual brain are (as we have seen in "Seeing the Man Through His Brain," page 33) as distinctive as a fingerprint, are found in almost identical numbers in everyone and have only relatively small variations (on the order of only a centimeter or two—well under an inch) in their positions.

To identify common structural features among brains more precisely, the concept of an "average" brain has been developed.

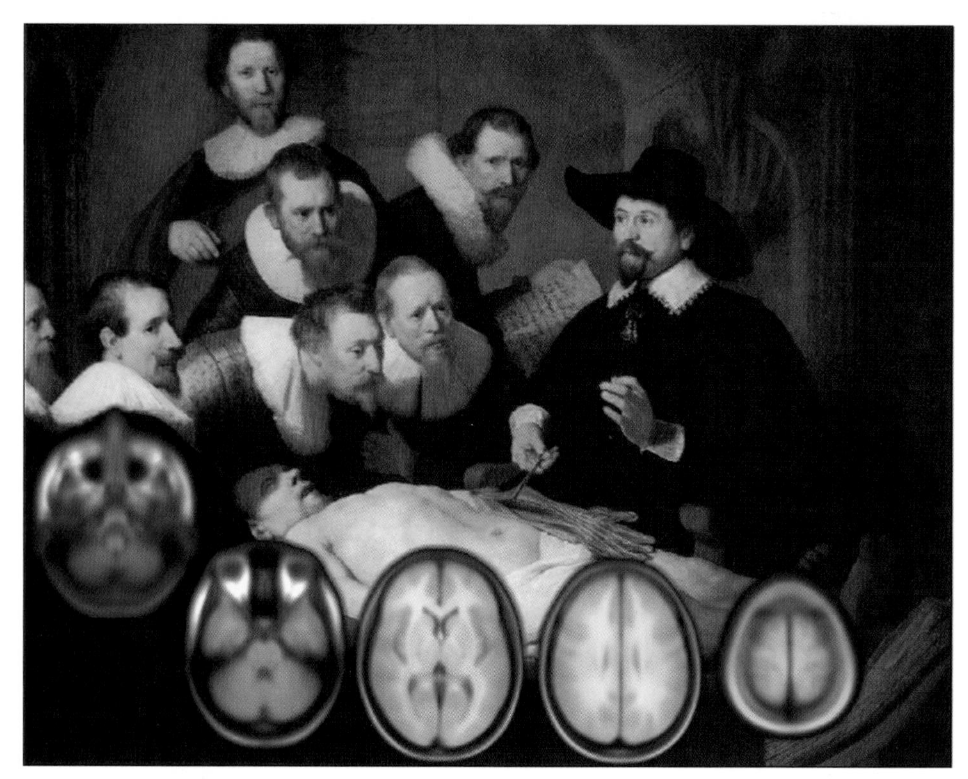

Figure 7. *The availability of high-resolution structural imaging data on large numbers of individual brains has led to a "new anatomy." This is based on defining a "probabilistic" outline of the brain in a common brain space. Probabilistic in this case means that the images shown here do not define structures in any individual but give an estimate of the likelihood of finding any particular brain structure at a given point in the whole population of subjects studied. This particular example is based on MRI scans from 305 normal individuals. In comparing this average brain map for a group with the individual brains shown in Figure 6, note that details of the foldings of the brain are not apparent in the average images in many areas where there is considerable variation among individuals.*

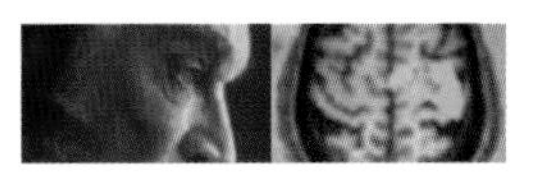

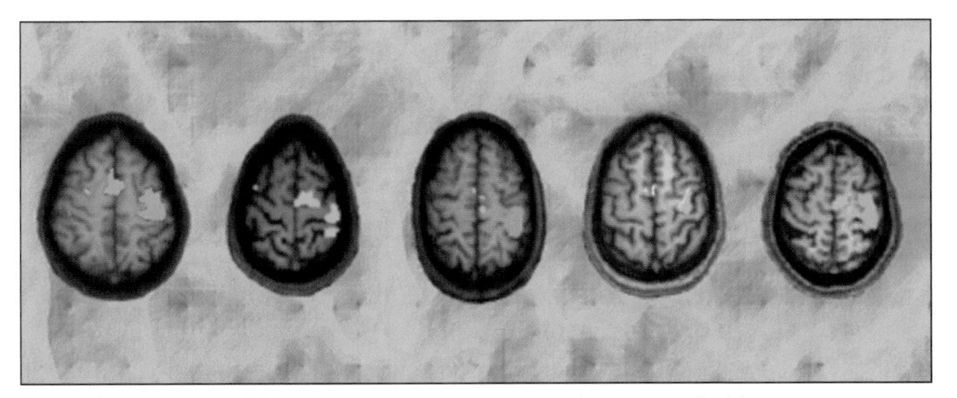

Figure 8. *Patterns of functional brain activation show remarkable similarity among individuals. Here, fMRI images from six individuals who all performed a simple finger-tapping task are superimposed on each other. The areas of brightest color are areas of activation in the greatest number of individuals. Large areas over the regions of the brain known to be involved in motor tasks are activated in common by all of the subjects, confirming that common brain mechanisms are used to perform this task.*

A representation of such an average brain would encompass descriptions of Shylock's brain as well as Antonio's, Bassanio's, Portia's, and even our own. One such average brain representation is illustrated here (Fig. 7). To make this picture, Alan Evans and his colleagues at the Montreal Neurological Institute used MRI scans of the brains of 305 normal individuals. With the aid of computer methods, they then stretched or shrank the individual brains (with some twisting) until they best fitted into a common shape. The image here shows the likelihood of finding a particular structural feature of the brain at any point in the so-called common brain space. Those features that are most sharply defined occupy the same positions in all individuals, but those that are not (for example, the gyri in the front of the brain) show greater variability.

Studies like this show that all human brains share the same overall structure, with only minor variations. But does that mean all human brains work in the same way? If different normal individuals perform the same tasks, will the same parts of their brains be activated? The development of functional imaging methods that can map brain activity is now making it possible to answer such questions. This is illustrated in Figure 8, on the previous page, in which we see composite images produced by fMRI of patterns of brain activation in several normal volunteers who were given the simple task of tapping the fingers of one hand. Some small differences probably arise from recording noise in the images or variations in the ways in which individuals perform the task. What is most striking is that the areas of activation in the brains are so similar. The right side of the brain always controls movement on the left side of the body and vice versa, for example. Thus, the right hand can move only after neuronal activation in a specific region (the primary sensorimotor cortex) that borders the large central sulcus near the top of the brain in the left hemisphere. Hand tapping is also always associated with another region of activation, near the midline at the top of the brain, where movement planning takes place. Perhaps even more remarkable is that patterns of activation for more complex cognitive functions, such as reading a word or recognizing a pleasant taste, also show great similarities among individuals. This suggests that all human brains share a common organization of functions.

Given all of these similarities, how is it that individual differences arise? Why are some people smarter, more creative, or just different? Why is it that Shylock reacts to his social exclusion by choosing revenge, while Romeo feels no bitterness but only longing for his Juliet after his banishment from Verona? The answers to these important questions are not yet known. However, it is clear that learning, skill development, memory, and many other cognitive processes critical to making each of us unique depend strongly on dynamic changes in the interactions between different areas of the brain. Differences in the patterns and strengths of specific interactions in the brain that are critical in determining individual differences are a function of our genes, as well as environment and experience. Our common humanity arises from a shared structure and shared mechanisms for emotions and behavior in the brain.

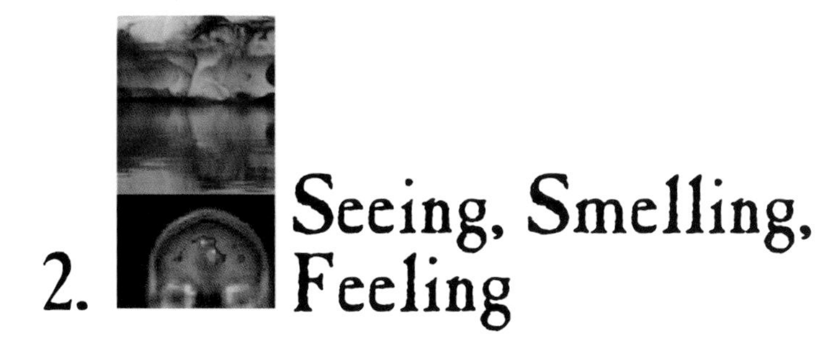

2. Seeing, Smelling, Feeling

FINDING A FACE

Love at first sight takes center stage in Romeo and Juliet. *Romeo, the young son of the wealthy Montagues, first meets the beautiful young Juliet, daughter of his family's archrivals, the Capulets, in Verona at a feast in the Capulet household. Their love is based initially on simple physical attraction. After Romeo and his friends leave the Capulet household, Romeo steals back to the garden beneath Juliet's bedroom window, hoping to catch another glimpse of her. When Romeo sees Juliet's face at her chamber window, he instantly recognizes his newfound love. He concentrates on her face, because it is through expressions of the face and eyes that we recognize and communicate emotion.*

Romeo: But soft, what light through yonder window breaks?
It is the east and Juliet is the sun!
Arise fair sun and kill the envious moon
Who is already sick and pale with grief
That thou her maid art far more fair than she.

Romeo and Juliet.
Alexandra Gilbreath as Juliet and David Tennant as Romeo.

Be not her maid since she is envious,
Her vestal livery is but sick and green
And none but fools do wear it. Cast it off.
It is my lady, O it is my love!
O that she knew she were!
She speaks, yet she says nothing. What of that?
Her eye discourses, I will answer it.
I am too bold. 'Tis not to me she speaks.
Two of the fairest stars in all the heaven,
Having some business, do entreat her eyes
To twinkle in their spheres till they return.
What if her eyes were there, they in her head?
The brightness of her cheek would shame those stars
As daylight doth a lamp. Her eyes in heaven
Would through the airy region stream so bright
That birds would sing and think it were not night.
See how she leans her cheek upon her hand.
O that I were a glove upon that hand,
That I might touch that cheek.

Romeo and Juliet, 2.2

Her eye discourses, I will answer it.

Shakespeare's plays are filled with descriptions of sensations. *Romeo and Juliet* is particularly rich in these, as befits a tale of young love. Of the five senses (vision, hearing, touch, smell, and taste), the greatest source of information about the world for most of us is provided by vision. There are many very vivid descriptions of the visual world in Shakespeare's plays, particularly in word paintings of the dramatic light of early morning or night, as in this beautiful moonlit scene.

Our biological evolution allows visual information to speak to us very directly in art. Our cultural evolution has further emphasized the importance of visual art. The power of such art—the Acropolis, the Mona Lisa, the brilliant colors of a Matisse—relies on the ability of the visual systems of the brain to link quite directly into memory to evoke associations that give the forms meaning. Interactions of both vision and memory with the emotional centers of the brain generate the pleasure that is part of our sense of aesthetics.

When we talk of "seeing," we are not speaking of eyes at all: we are speaking of an activity of the brain. The brain activity in "seeing" occurs specifically in the primary visual cortex, which is found in a thin rim of gray matter at the back of the brain along the inner surface of each hemisphere. As in other areas of the brain, the neurons in the visual cortex and their organization are highly specialized. In principle, if ways could be found to link the visual cortex to a computer, people who are blind because of advanced eye disease might have their vision restored by means of digitized video-camera images—a form of neural prosthesis already in experimental development.

A striking demonstration of how the brain processes visual images is provided by fMRI (Fig. 9). To generate the figure here, visual stimuli were shown to the subject's eyes alternately. Brain activation associated with stimulation of one eye is illustrated here in blue, and that from stimulation of the other eye in red. Note the alternating bands of activation caused by the stimulation of the two eyes. Each eye dominates the input in distinct regions of the primary visual cortex. This alternating pattern of "ocular dominance" in the visual cortex is set early in development. If computer "inflation" of the images is used to remove the surface folds of the brain, the borders between areas of the brain responding to each eye are revealed particularly clearly.

How is an image as complex as the scene viewed by Romeo processed? Scientists are still working to understand this fully, but the Nobel Prize-winning work of David Hubel and Torsten Wiesel (shared with Roger Sperry in 1981) helped initiate our current view, showing that certain cells in the visual system respond to specific simple features, such as the direction of lines, while other cells respond to more complex features. This led to the notion that information travels through multiple levels of hierarchical processing in order to build up a "complete" image. Imagine, for example, the problem of seeing a house.

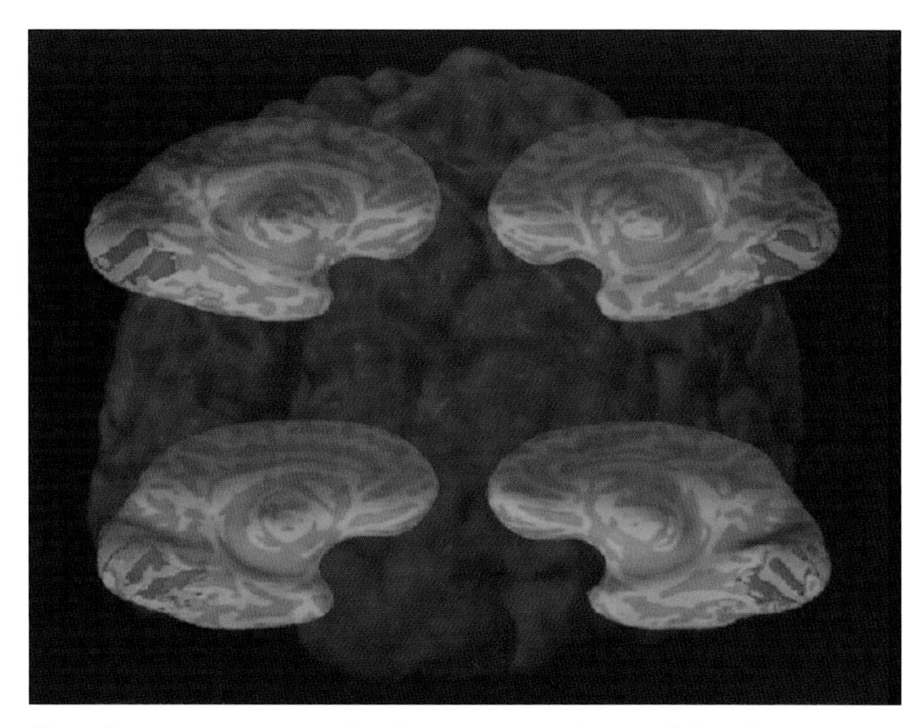

Figure 9. *Brain imaging studies have increasingly provided evidence that specific areas of the brain are activated in processing special types of information, such as the visual recognition of faces. This image shows fMRI activations superimposed on a brain map generated by "inflating" the normal brain to flatten out the folds. The red and blue areas show parts of the brain that become active with stimulation of either the right or the left side of the field of vision. In contrast, the green area represents activity of the brain that occurs specifically when a face is viewed. This occurs in a part of the brain known as the fusiform gyrus.*

First, one must differentiate areas of house color from the background of the sky, then one must identify the edges of the house and individual features such as the door, the windows, the roof, and so on. Critical to distinguishing a ruin from a welcoming home is the relative orientation of the different edges. Finally, the relative positions, colors, and textures of the different parts must be integrated within the brain's picture.

Modern imaging has helped reveal the parts of the brain that process visual information. One of the most interesting of these special regions is associated with the recognition of faces. Recognizing faces is critical for us, because it is largely by the face that we distinguish friend from foe or, as with Romeo, select a mate. Facial expressions also provide an important medium for communication, particularly of emotional experience.

In the second part of the experiment whose results are illustrated in Figure 9, the scientists presented pictures of faces to their subjects during the fMRI scanning. Specific areas found on the lower surface of the brain were shown to become active just when viewing faces. Injury to these areas is rare but can give rise to extraordinary clinical phenomena. One such patient was famously described by the neurologist and author Oliver Sacks as "the man who mistook his wife for a hat" and was immortalized in his book of that name. This unfortunate patient had suffered strokes in the back of his brain that affected regions responsible for the integration of visual forms. While he was not blind, the patient could not mentally assemble features that he perceived visually into a coherent whole without great effort. This could lead to considerable confusion, such as mistaking his wife for a hat!

SMELL: A DIRECT LINK TO THE EMOTIONAL BRAIN

Shakespeare uses the sense of smell metaphorically in Hamlet. *The young Prince of Denmark is not the only character in the tragedy to deliver a moving soliloquy; King Claudius, his treacherous uncle, also speaks his thoughts aloud, describing his sin using the powerful metaphor of smell. Claudius has become king by poisoning his own brother (Hamlet's father) and marrying his sister-in-law, Gertrude (Hamlet's mother). Believing himself beyond forgiveness, Claudius kneels and tries to pray for forgiveness. Hamlet enters as Claudius kneels. Although ready to kill his uncle, Hamlet decides not to do so, for he fears that if Claudius dies while praying, his soul might go to heaven. The irony, of which only the audience knows, is that Claudius rises soon afterward, unable even to begin a prayer.*

King: O, my offence is rank, it smells to heaven;
It hath the primal eldest curse upon't—
A brother's murder. Pray can I not,
Though inclination be as sharp as will,
My stronger guilt defeats my strong intent,
And, like a man to double business bound,
I stand in pause where I shall first begin,
And both neglect. What if this cursed hand
Were thicker than itself with brother's blood,
Is there not rain enough in the sweet heavens
To wash it white as snow? Whereto serves mercy
But to confront the visage of offence?
And what's in prayer but this twofold force,
To be forestalled ere we come to fall
Or pardon'd being down? Then I'll look up.
My fault is past—but O, what form of prayer

Hamlet. Sam Waterston as Hamlet and Charles Cioffi as Claudius.

Can serve my turn? 'Forgive me my foul murder?'
That cannot be, since I am still possess'd
Of those effects for which I did the murder—
My crown, mine own ambition, and my queen.
May one be pardon'd and retain th'offence?
In the corrupted currents of this world
Offence's gilded hand may shove by justice,
And oft 'tis seen the wicked prize itself
Buys out the law. But 'tis not so above:
There is no shuffling, there the action lies
In his true nature, and we ourselves compell'd
Even to the teeth and forehead of our faults
To give in evidence. What then? What rests?
Try what repentance can. What can it not?
Yet what can it, when one can not repent?
O wretched state! O bosom black as death!
O limed soul, that struggling to be free
Art more engag'd! Help, angels! Make assay.
Bow, stubborn knees; and heart with strings of steel,
Be soft as sinews of the new-born babe.
All may be well.

Hamlet, 3.3

O, my offence is rank, it smells to heaven;
It hath the primal eldest curse upon't—
A brother's murder.

In this passage, Claudius reflects on the possibility that he may have gained the crown of Denmark only to lose his soul: "In the corrupted currents of this world/Offence's gilded hand may shove by justice,/. . . but 'tis not so above." Despite his overweening ambition and the cold-blooded murder of his brother, Claudius has a moral sense, and he is genuinely frightened by the prospect of eternal damnation. He powerfully conveys his appreciation for the evil he has done as he cries, "O, my offence is rank, it smells to heaven."

The metaphor of a noxious odor is a moving way to convey the repugnance of a morally foul act. Although vision dominates so much of our conscious sensory experience, smell is perhaps the most sensitive of all our senses. Some odorants can be detected down to concentrations of even a few parts per million, allowing responses to be made to even small numbers of molecules. Just so, it is possible for us to recognize a single evil act, even if all the other acts that we see a person perform are good.

We can distinguish an enormous range of scents. Estimates suggest that odor receptors in the soft tissue high inside the nose are sensitive to 10,000 or more different odors. More remarkable, we can appreciate novel blends of odors while being able to identify the individual odors that contribute to these blends. Similarly, we are often sensitive to malice even when it is disguised. In the opening scenes of *Hamlet*, Claudius solicitously inquires after the cause of Hamlet's melancholy, yet we appreciate undercurrents of menace in his words. Claudius wants Hamlet to stop mourning and to celebrate his mother's marriage, even though it has followed a mere two months after his father's death. The dramatic irony of the scene lies in our knowledge that Claudius is motivated more by anxiety over Hamlet's intentions as heir to the throne than by concern for his well-being.

Claudius believes that the stench of his evil deeds will be recognized easily in heaven no matter how he tries to disguise them by *shuffling*, a word that here refers to fine words and outwardly generous behavior. More important, he believes that punishment in the afterlife will be rapid and unequivocal. This adds to the appropriateness of the metaphor. Unlike sensations of vision and touch, which proceed through intermediate processing areas deep in the brain, sensations from smell are targeted directly to parts of the brain responsible for emotion, the so-called limbic system, a group of structures found predominantly along the midline of the brain. Emotional responses to odors are thus elicited very directly.

Evolution must have selected strongly for a sophisticated sense of smell, just as our cultural heritage has selected for development of a moral sense. Just think how important it was for our remote ancestors to be able to guess immediately, from

only a faint smell, that food or water might be unsafe to consume. Smell probably contributed to mating choices, as well.

Understanding how systems for taste and smell are organized has been the goal of a number of recent functional imaging studies (Fig. 10). Exposure to a strong odor increases activity at the base of the front of the brain, in an area over the eyes called the orbitofrontal cortex. Other research has identified this same region as one that becomes more active in response to sensations that have strong hedonic qualities (that is to say,

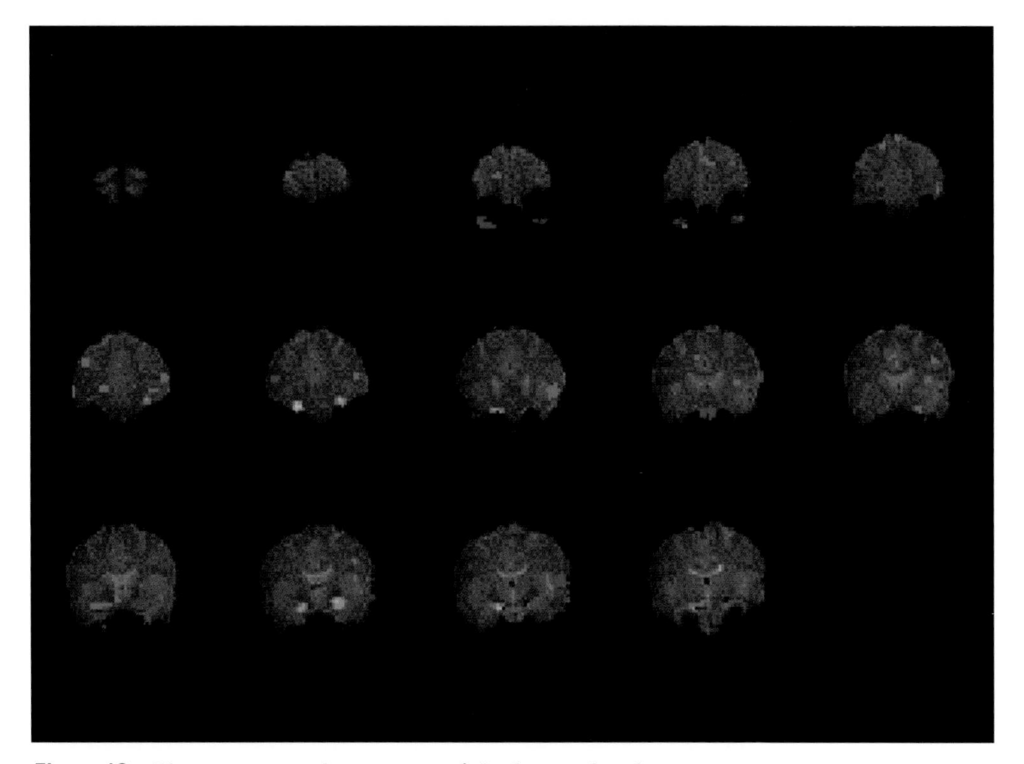

Figure 10. *These images show areas of the brain that become active with smell. The subject was studied by fMRI as brief bursts of scent were delivered. Many areas of the brain were activated together in experiencing an odor. The pathways in the brain responsible for processing the sense of smell are distinct from those involved in other sensory systems, such as vision and touch.*

marked degrees of pleasantness or unpleasantness). The orbitofrontal cortex has connections to the amygdala, an almond-shaped cluster of neurons in the middle of the brain. The amygdala is a key structure in the limbic system and is important for generating an appropriate response to the emotional qualities of sensations. Research suggests that unpleasant odors that give rise to a strong emotional response (as experienced by Claudius) excite the amygdala.

Other imaging studies have demonstrated that the brain can respond to odors that are so faint we are not even conscious of smelling them. The almost vanishingly faint odors of the natural sexual attractants known as pheromones, for example, can induce responses at a subliminal level. It is interesting to reflect that the sensory world of many animals must be dominated by smell as much as our sensory world is dominated by vision. Anyone who has taken a young puppy for a walk will be able to imagine the rich variety and blinding brightness of smells that the dog experiences in the out-of-doors. Smell in animals is a strongly motivating sensory cue. While we may sometimes prefer to ignore this aspect of our biological heritage, it is very much a part of some of our most important behaviors. Sales of expensive perfumes and aftershaves are driven by the notion that scent can be a powerful motivator for sexual attractiveness, for example—even if an unconscious one.

HEAT AND COLD: AS MUCH IN THE MIND AS ON THE SKIN

Seeing, Smelling, Feeling

Human reactions to heat and cold, Shakespeare suggests, may involve states of mind as much as changes in temperature. Just before its violent climax, Hamlet *offers a final comic scene. Hamlet is conversing with his friend Horatio when they are interrupted by the courtier Osric. Hamlet wants to embarrass Osric and to reinforce the prevailing notion that he has himself gone mad. Hamlet thus shifts unpredictably in talking about the weather, insisting alternately that he feels too hot or too cold. Osric tries to agree with both positions, proving himself a sycophantic fool.*

Osric: Sweet lord, if your lordship were at leisure, I should impart a thing to you from his Majesty.

Hamlet: I will receive it, sir, with all diligence of spirit. Your bonnet to his right use: 'tis for the head.

Osric: I thank your lordship, it is very hot.

Hamlet: No, believe me, 'tis very cold, the wind is northerly.

Osric: It is indifferent cold, my lord, indeed.

Hamlet: But yet methinks it is very sultry and hot for my complexion.

Osric: Exceedingly, my lord, it is very sultry—as 'twere—I cannot tell how. My lord, his Majesty bade me signify to you that he has laid a great wager on your head. Sir, this is the matter—

Hamlet. Tom Hulce as Hamlet, Paul Mullins as Rosencrantz, and J.C. Cutler as Guildenstern.

Hamlet: [*signing to him to put on his hat*] I beseech you remember—

Osric: Nay, good my lord, for my ease, in good faith.
Sir, here is newly come to court Laertes—believe
me, an absolute gentleman, full of most excellent
differences, of very soft society and great showing.
Indeed, to speak feelingly of him, he is the card or
calendar of gentry; for you shall find in him the
continent of what part a gentleman would see.

Hamlet, 5.2

I thank your lordship, it is very hot.

No, believe me, 'tis very cold, the wind is northerly.

It is indifferent cold, my lord, indeed.

After Osric tells Hamlet that the king has bet on him in a fencing match against Laertes, Hamlet accepts the challenge to fight. For the audience this scene is both comical and darkly menacing. The menace lies in the audience's knowledge, from a prior scene, that Claudius and Laertes are plotting to kill Hamlet during this outwardly friendly match. Yet it is impossible not to laugh at Osric's bumbling efforts to agree with Hamlet's shifting and contradictory remarks about the weather.

The wordplay heightens the dramatic tension. It also illustrates a fundamental issue concerning sensations. Shakespeare knows that his audience will believe that the sensations of heat and cold are distinct. The body is able to distinguish a wide variety of sensations, including most particularly those of temperature, touch, joint position, and vibration, because the different receptors in the skin are sensitive to distinct types of stimuli. Signals from these receptors are arranged in separate, parallel circuits that run to the brain. The circuits are built up from axons (the nerve fibers that make up the wiring of the central nervous system) having different sizes and varying abilities to conduct signals. For example, the axons carrying temperature sensations are very thin and conduct signals particularly slowly (that is why there is a detectable delay between putting a foot in an overly hot bath and withdrawing it in pain). In contrast, the axons that carry information about the position of our joints are thicker and conduct signals quickly, allowing us to make split-second adjustments of position and balance as we walk or meet the unpredictable bounce of a tennis ball with our racket.

Nonetheless, in this exchange between Hamlet and Osric, Shakespeare comically reminds us that our final perception of hot and cold is a subjective experience—the sensation is what our brain tells us it is! Our skin temperature sensors better measure relative than absolute temperature. (Try warming the right hand over the stove while holding ice in the left, then plunging both hands in a sink of tepid water—the right hand will sense the water as cold, while to the left it will feel hot.) Extremes of heat and cold can also be confused. Just as we can be burned by a hot stove, so the prolonged touch of dry ice or another very cold object can "burn." The overwhelming sensation with either is of a type of pain rather than hotness or coldness.

These apparent paradoxes are due in part to limitations of the temperature receptors in the skin and in part to the fact that the brain has to "interpret" their sensory signals. Sensations of extreme heat or cold also stimulate a special set of receptors for noxious stimuli in the skin. Signals from these so-called nociceptive receptors are transmitted to the brain in parallel with signals from the specialized temperature discrimination receptors. In the brain they are processed together by cells in a structure near the center of the brain known as the thalamus. The thalamus is connected to a wide range of other brain areas and acts as an integration center for sensory input. The primary sensory areas of the brain then become activated, giving rise to the conscious experience of temperature and allowing us to locate its source in a particular part of the body.

However, recent imaging studies of the brain have emphasized that a dominant part of the brain activity triggered by extremes of temperature is concerned with the associated emotional response ("That hurts!") rather than the specifics of heat or cold (Fig. 11). Irene Tracey of the University of Oxford, working with colleagues at the Massachusetts General

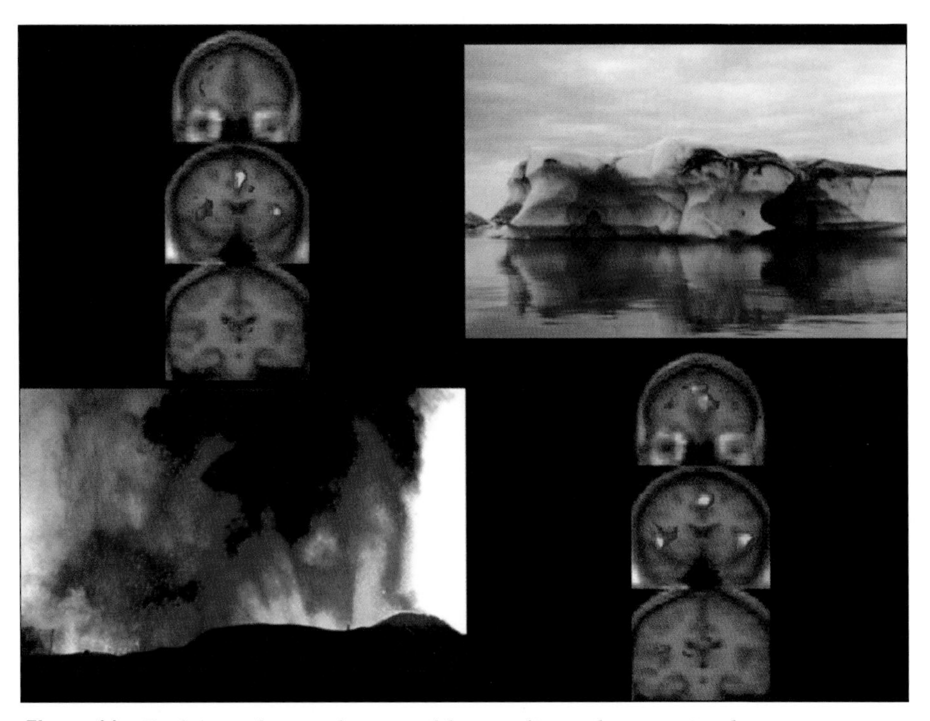

Figure 11. *Both very hot and very cold stimuli can be perceived as painful. The fMRI images here illustrate patterns of brain activity associated with either a painfully hot or a painfully cold stimulus applied to the hand. The responses to the pain of both stimuli involve many areas of the brain, particularly areas in the limbic system (the so-called emotional brain). Activation of these areas is responsible for the sense of unpleasantness associated with both painful stimuli. The striking similarity of brain activity associated with these two very different stimuli suggests that the common element of unpleasantness contributes more to the activation of the brain during the stimuli than the characteristics of hotness or coldness.*

Hospital NMR Center, studied the response of subjects to extremes of heat and cold delivered to the hand. She found that the patterns of brain activation produced by both extremes are almost identical, as shown in the image here, in which selected parts of the brain that are activated in response to heat stimuli are represented above the image of a volcano, while those activated in response to cold stimuli are shown below the picture of an ice floe.

This "logic" of the brain seems to make good sense: the most important response to a noxious stimulus, whether it is too hot or too cold, is to move away quickly. Only later does it become important to worry about the qualities of the stimulus and what part of the body it has affected. Because perceptions of extreme heat and cold, as well as other noxious stimuli, are purely mental phenomena, it is possible to train oneself to ignore them, using meditation and other psychological techniques. This helps explain how Hindu ascetics who perform feats of magic and endurance are able to walk on hot coals or lie on beds of nails without howling in pain, while most of us would cringe at the very thought of doing such things!

3. 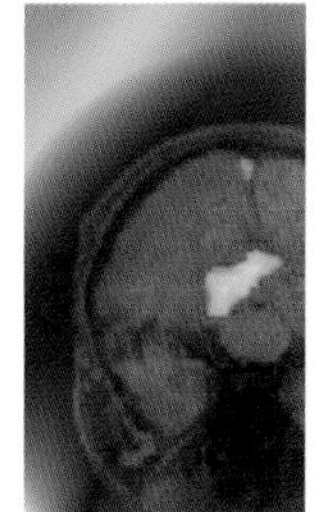 Decision and Action

BATTLE OF THE SEXES

In Shakespeare's day there was no scientific evidence for differences between men's and women's brains, other than perhaps that women's brains tend to be slightly smaller than those of men (an anatomical fact that does not mean that men must be smarter!). There was, however, the evidence of generations to attest to the seemingly irreconcilable differences that fuel the conflicts and account for at least some of the mutual appeal of the sexes. In The Taming of the Shrew *Shakespeare makes some of his greatest comedy out of the puns or double meanings that afford a linguistic metaphor for the duality of the sexes. When the impoverished gentleman Petruchio comes to Padua to find a wife, he hears of a wealthy man's unruly daughter and decides to court the spirited Katherina, widely known as a shrew because of her violent temper. Petruchio woos her with characteristic good humor, teasing her with the nickname "Kate" (a play on the word* cate, *meaning a special or choice food). In the scene that follows tempers flare, passions collide, and the battle of the sexes is joined afresh.*

Petruchio: Good morrow, Kate, for that's your name, I hear.

Katherina: Well have you heard, but something hard of hearing:
They call me Katherine that do talk of me.

Petruchio: You lie, in faith, for you are call'd plain Kate,
And bonny Kate, and sometimes Kate the curst;
But Kate, the prettiest Kate in Christendom,
Kate of Kate Hall, my super-dainty Kate,
For dainties are all Kates, and therefore, Kate,
Take this of me, Kate of my consolation,
Hearing thy mildness prais'd in every town,
Thy virtues spoke of, and thy beauty sounded,
Yet not so deeply as to thee belongs,
Myself am mov'd to woo thee for my wife.

Katherina: Mov'd, in good time! Let him that mov'd you hither
Remove you hence. I knew you at the first
You were a movable.

Petruchio: Why, what's a movable?

Katherina: A joint-stool.

Petruchio: Thou hast hit it. Come, sit on me.

Katherina: Asses are made to bear, and so are you.

Petruchio: Women are made to bear, and so are you.

Katherina: No such jade as you, if me you mean.

Petruchio: Alas, good Kate, I will not burden thee!
For, knowing thee to be but young and light—

Katherina: Too light for such a swain as you to catch,
And yet as heavy as my weight should be.

The Taming of the Shrew, 2.1

The idea that "men are from Mars, women from Venus" is hardly a new one. Every generation rediscovers the fundamental differences between men and women. Petruchio describes

The Taming of the Shrew. Tracey Ullman as Katherina and Morgan Freeman as Petruchio.

Women are
made to bear,
and so are you.

No such jade as
you, if me you
mean.

the contradictions of "this wonderful woman Katherine/Kate" when he asks whether she is "bonny Kate" or "Kate the curst." There is certainly great irony when he speaks of Katherina's "mildness prais'd in every town." Thus, as he proceeds to woo her he also wants to "tame" her. But the subjugation of love involves reaching a mutual understanding, for both Katherina and Petruchio want to have their own way—at least at the onset of the play.

Katherina is as quick a wit as the fiery-tongued Beatrice in *Much Ado About Nothing*. She responds with sharp wordplay to Petruchio's teasing. He tries to get the better of her, but in this scene at least she ends up one step ahead. Arguing that Katherina should be ready for marriage and children, Petruchio reminds her, "Women are made to bear, and so are you." Katherina, quickly seizing on a different meaning, replies, "No such jade as you, if me you mean," emphasizing her unwillingness to "bear" (or tolerate) Petruchio. Petruchio, acknowledging her brief victory, tactically withdraws from the verbal skirmish: "Alas, good Kate, I will not burden thee!"

The intriguing differences between men and women have set modern brain scientists to work to search for differences in cognitive processing between the sexes. In part, any such differences could be due to sex-linked influences on the development of the brain. They also could be due to additional hormonal influences on brain function later in life. Of course, cultural differences in the way girls and boys are raised, and men and women are treated, may contribute as well.

Figure 12 illustrates one approach to understanding the ways in which male and female brains may differ in function. Earlier psychological studies had suggested that men and women respond differently when they are shown pictures of a pair of irregular objects and asked whether they are rotated views of the same object. Women tend to respond faster in this "mental rotation" task than men, but men tend to be more accurate. To do the task, a person must mentally rotate one of

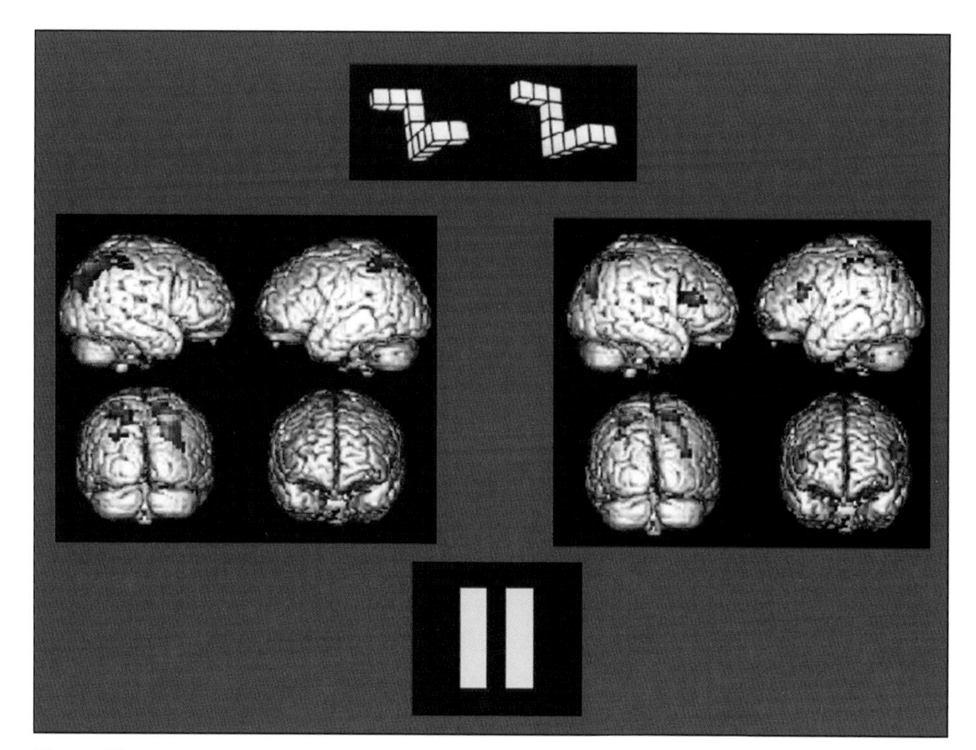

Figure 12. *Researchers found that men and women showed differences in the rate and accuracy with which they solved a spatial problem. In contrasting the patterns of brain activation between the sexes, they found that the patterns of fMRI activation were significantly different during a task demanding a decision as to whether the two figures displayed (see top of picture) represented simple rotated forms of the same structure. Men (images to the left) tended to show more equal activation in the two sides of the brain than women. Women (images to the right) showed more activation in the front of the brain and around areas in the middle than did men.*

the objects to determine whether it can be made to fit precisely on the outline of the other.

The pattern of increased brain activity during this task is shown to the left for men and to the right for women. In both sexes there is increased activity in the front of the brain, where planning for movement occurs, as well as in areas responsible for remembering things over short periods of time (for example, the series of moves needed to complete a puzzle). In addition, there is activation in the back of the brain, in the so-called parietal lobes, where information on spatial relationships is believed to be integrated.

However, women show more activation in the front of the brain and in areas responsible for movement. Men show more activation in the back of the brain (parietal lobes) where spatial relations are computed. Although in this experiment (which involved small numbers of subjects) there were no clear differences in performance between the two groups, the results suggest that the two sexes may call on somewhat different cognitive processing strategies to solve the same problem. For women, the strategy may emphasize the motor response, perhaps accounting for a general tendency to respond more quickly. For men, the strategy may emphasize the development of a more detailed mental image of the object, possibly accounting for the general tendency to give more accurate responses.

What this type of functional brain imaging does not address, however, is whether the differences in activation represent fundamental differences in the "hard-wiring" of the brain, or whether they merely represent sex-linked differences in strategies for solving this type of problem. Some observations suggest that there may be differences in the hard-wiring of male and female brains (such as the earlier age of language acquisition for girls and the tendency for greater early manual dexterity in boys). However, much needs to be learned in this challenging and sometimes controversial area of developmental neuroscience. Whatever the precise nature of the differences between the sexes, Shakespeare has identified a fundamental aspect of human experience with his tale of Katherina and Petruchio—namely, that the differences between men and women are less important to recognize primarily than the greater commonality of cognitive processes and experience.

MOVEMENT BEGINS IN THE MIND

Decision and Action

As every athlete knows, concentration and mental rehearsal can improve even physical performance. "Once more unto the breach, dear friends," cries Henry V in one of Shakespeare's most frequently quoted (and—of course—often misquoted) lines. The king, addressing his army in front of the French town of Harfleur, calls for courage before their attack on the walled town. Henry encourages his fighters to enter into the fray by painting a mental picture of the bold movements of a fierce tiger. The day's fight will lead to the capture of the town by King Henry's forces.

Henry V: Once more unto the breach, dear friends, once more,
Or close the wall up with our English dead.
In peace there's nothing so becomes a man
As modest stillness and humility;
But when the blast of war blows in our ears,
Then imitate the action of the tiger:
Stiffen the sinews, conjure up the blood,
Disguise fair nature with hard-favoured rage.
Then lend the eye a terrible aspect;
Let it pry through the portage of the head
Like the brass cannon; let the brow o'erwhelm it
As fearfully as doth a galled rock
O'erhang and jutty his confounded base,
Swilled with the wild and wasteful ocean.
Now set the teeth and stretch the nostril wide,
Hold hard the breath and bend up every spirit
To his full height. On, on, you noble English,
Whose blood is fet from fathers of war-proof,
Fathers that like so many Alexanders
Have in these parts from morn till even fought,
And sheathed their swords for lack of argument.
Dishonour not your mothers; now attest

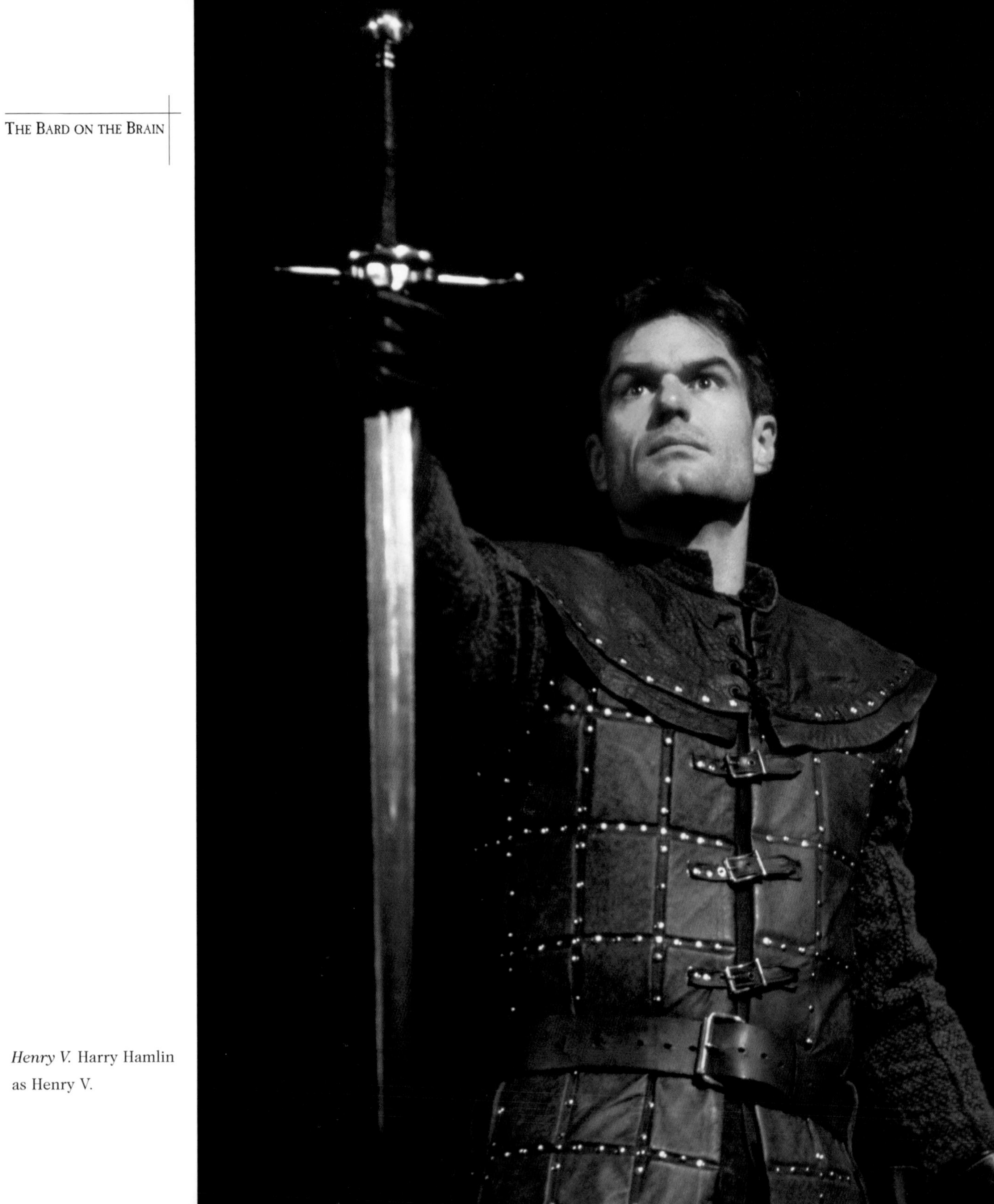

Henry V. Harry Hamlin as Henry V.

That those whom you called fathers did beget you.
Be copy now to men of grosser blood
And teach them how to war. And you, good yeomen,
Whose limbs were made in England, show us here
The mettle of your pasture; let us swear
That you are worth your breeding—which I doubt not,
For there is none of you so mean and base
That hath not noble lustre in your eyes.
I see you stand like greyhounds in the slips,
Straining upon the start. The game's afoot.
Follow your spirit, and upon this charge
Cry 'God for Harry! England and Saint George!'

Henry V, 3.1

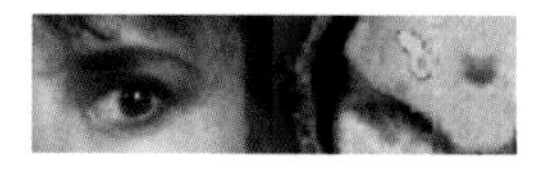

In this stirring speech, Henry rouses his troops to action, instilling in them a stronger sense of purpose by offering a greater cause to fight for: " . . . 'God for Harry! England and Saint George!'" Henry is also preparing them for action in more specific ways. In a society of rigid class distinctions, Henry addresses each social group in the army individually. He encourages a sense of entitlement in his nobles by reminding them of fathers that "like so many Alexanders/Have in these parts from morn till even fought." He goads them on, warning them: "Dishonour not your mothers." He addresses the common foot soldiers as sleek "greyhounds in the slips" and works to remove their fear by stressing that they should be eager for action ("Straining upon the start").

Now set the teeth and stretch the nostril wide, Hold hard the breath and bend up every spirit To his full height.

Henry's men are cold in the early-morning hours, tired from their march from the coast, and greatly outnumbered by the French. He knows that if his soldiers are slow to consolidate the small advantage they have in controlling the start of the battle, or sluggish in using their weapons, they will be injured or killed and that the loss of every man will be felt. To charge them for the battle to come and bring a unity of warlike purpose, Henry also speaks to the army as a whole.

He relies on rich visual imagery to inspire his nobles and the common men alike. Henry calls on them all to "imitate the action of the tiger:/Stiffen the sinews, conjure up the blood,/Disguise fair nature with hard-favoured rage." Each is asked to imagine making a bold and ferocious leap forward against the enemy walls. They must "conjure up the blood" not only to supply the metabolic needs of their muscles but also to motivate rage for their enemy so great that they do not hesitate to strike even for an instant. He urges them to " . . . set the teeth and stretch the nostril wide,/Hold hard the breath and bend up every spirit/To his full height." In essence, Henry employs a visualization technique not so different from that which an athletic coach might use today before a big game is to begin.

In this speech, Shakespeare recognizes a critically important aspect of any complex action: the need for planning. Usually we do not feel a conscious need to plan our movements, but even simple movements of a finger or a hand demand preparation. The target must be identified, the limb to be moved must be chosen, and then a trajectory must be calculated, as well as the individual movements that will follow. Remarkably, all of this can be done by the human brain in only hundreds of milliseconds. More complex movements may demand greater planning and longer preparation times. On the battlefield, differences in response times as small as fractions of a second can make the difference between wounding and being wounded. The movement planning inspired by Henry's imagery is calculated to help give his soldiers the critical edge needed to triumph over the French.

When a targeted movement such as the soldiers would use in raising their swords against the enemy is performed, many areas of the brain must work together. The pictures shown in Figure 13 illustrate four images of the brain of a person reaching toward a simple target while being monitored by fMRI. This experiment, performed by Chris Miall of Oxford University, shows very nicely the major areas of the brain that are engaged in such a task. In the upper middle part of the brain, there is

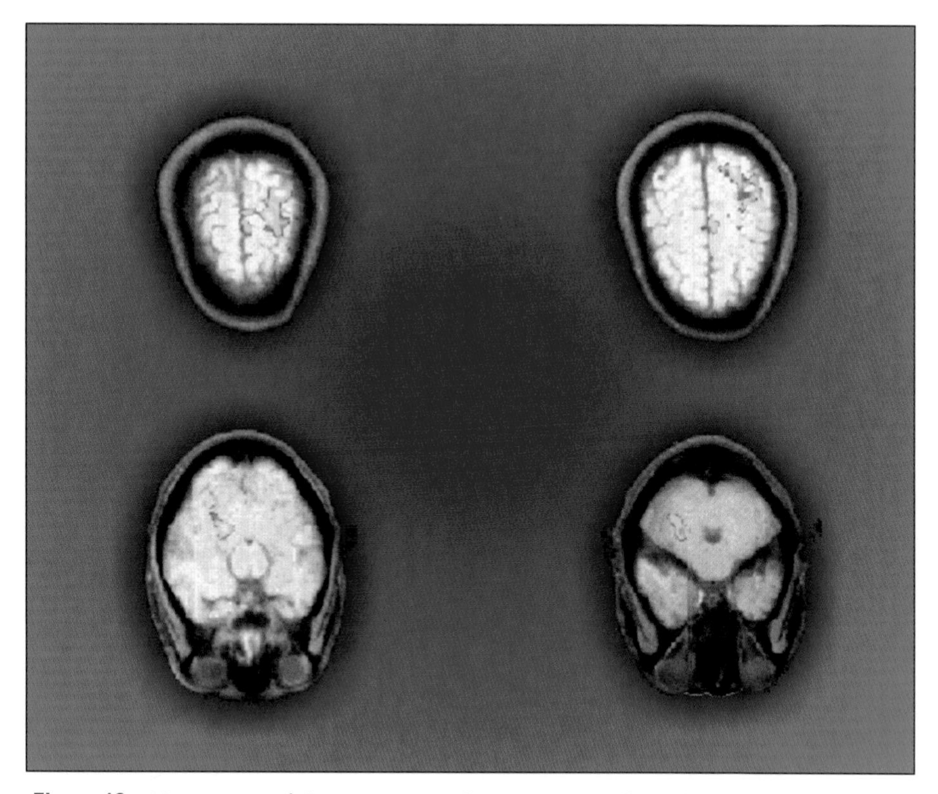

Figure 13. *Movement of the arm toward a target involves the integrated activity of the perceptual and motor planning and execution centers of the brain. Here the pattern of brain activation associated with a visually guided movement of the forearm toward a simple target is shown. The upper images show prominent activation in areas of the brain known to be involved with motor planning and execution (left) and the representation of the relative positions in space of the start and the end of a movement (right). The image on the lower right shows activation in the cerebellum, or "hindbrain," during this task. Injury to the cerebellum can cause a tremor and impair the accuracy of movement toward targets.*

activation of a key planning center for movement (the supplementary motor area). On the side of the hemisphere opposite to the hand being moved, there is activation of the specialized part of the brain (the primary sensory cortex) that responds to sensory information from the limb. We are conscious of only a fraction of this. While we are consciously aware of sensory information related to "touch" on a limb, we process entirely unconsciously the considerable feedback from nerves in joints and muscles that is necessary for the limb's position in space and its speed to be adjusted by the brain during a movement. Immediately in front of the primary sensory cortex (and not distinguishable from it in this experiment) is the primary motor area, which sends the command for movement to the limb. Coordination of the movement is facilitated by activity in the "hindbrain," or cerebellum, which is shown in the image in the lower right-hand corner. In the cerebellum, most of the activation associated with movement occurs on the same side as the limb being moved.

Other work has clearly demonstrated that activity in motor planning areas occurs well before the primary motor cortex sends the signal for movement. Activation of these planning areas, farther toward the front of the brain, can be identified separately in experiments in which the movement itself is delayed after the signal to move is given. Functional imaging studies have also shown how activity in motor planning areas increases if a movement is simply imagined to occur. As imagining a movement can increase the speed with which a movement is performed, this activation of the movement planning area is likely to provide a training effect similar to that of movement itself. It is this phenomenon that Henry exploits at the siege of Harfleur.

"LET ME CLUTCH THEE"

Decision and Action

The brain must control every motor skill—even one as simple as closing the hand to grasp an object. After plotting to kill the king, Macbeth is shocked by the vision of a bloody dagger. He tries to take hold of it but fails. He attributes this "fatal vision" of the murder weapon to a "heat-oppressed brain." A bell interrupts the soliloquy, and Macbeth leaves to commit his crime.

Macbeth: Is this a dagger, which I see before me,
The handle toward my hand? Come, let me clutch thee:—
I have thee not, and yet I see thee still.
Art thou not, fatal vision, sensible
To feeling, as to sight? or art thou but
A dagger of the mind, a false creation,
Proceeding from the heat-oppressed brain?
I see thee yet, in form as palpable
As this which now I draw.
Thou marshall'st me the way that I was going;
And such an instrument I was to use.—
Mine eyes are made the fools o'th' other senses,
Or else worth all the rest: I see thee still;
And on thy blade, and dudgeon, gouts of blood,
Which was not so before.—There's no such thing.
It is the bloody business which informs
Thus to mine eyes.—Now o'er the one half-world
Nature seems dead, and wicked dreams abuse
The curtain'd sleep: Witchcraft celebrates
Pale Hecate's off'rings; and wither'd Murther,
Alarum'd by his sentinel, the wolf,
Whose howl's his watch, thus with his stealthy pace,
With Tarquin's ravishing strides, towards his design
Moves like a ghost.—Thou sure and firm-set earth,
Hear not my steps, which way they walk, for fear

Macbeth. Stacy Keach as Macbeth.

Thy very stones prate of my where-about,
And take the present horror from the time,
Which now suits with it.—Whiles I threat, he lives:
Words to the heat of deeds too cold breath gives.

[*A bell rings*.]

I go, and it is done: the bell invites me.
Hear it not, Duncan; for it is a knell
That summons thee to Heaven, or to Hell.

Macbeth, 2.1

This passage powerfully conveys Macbeth's compulsion to clutch the spectral dagger that appears to him. Like the audience, he knows the dagger is an illusion. Nonetheless, he is unable to suppress the urge, so he grasps the "palpable" dagger that hangs from his warrior's belt.

Although this situation is extraordinary, it is an illustration of what each of us does scores of times each day: reaching and grasping. These are complex motor acts that involve planning—unconsciously (as described in the previous section) or consciously. You may be unaware of doing it, but before you reach to turn the page of this book, you will have identified the location of the corner of the page, plotted a trajectory, decided the required velocity and force, and planned how to open and close your fingers to grasp it. A golfer trying to improve his stroke may perform similar tasks very deliberately indeed, taking the time to gauge the weight of the club, the direction and speed of the wind, and the relative results of slightly different types of stroke.

Is this a dagger, which I see before me, The handle toward my hand? Come, let me clutch thee:—I have thee not, and yet I see thee still.

As described in the previous section, precise movement toward a target involves many areas of the brain. But clinical observations show that critically important activity occurs in the cerebellum. When the cerebellum (or, more precisely, the side, or cerebellar hemisphere) is damaged, patients can develop a pronounced shaking (tremor) on that side when moving the arm. The shaking typically gets worse as the target is approached. This makes it appear as though the brain is continually trying to correct the path of movement but always overadjusting in doing so. The movements are clumsy, or ataxic, and lose their smoothness. These deficits demonstrate that the cerebellum normally helps coordinate activities of other parts of the brain.

Figure 14. *To generate these fMRI brain activation images, a subject was asked to reach toward and then grasp a small ball held before him. The images show prominent activation in the cerebellum, or "hindbrain." Different areas in the cerebellum become excited with movements of different parts of the body. These areas are organized according to the outward form of the body. The greatest activation of the cerebellum is on the same side of the body as the arm being moved (in contrast with activation of the motor cortex, which shows greatest activation in the hemisphere opposite to the hand being moved).*

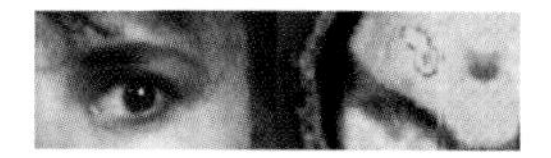

The substantial involvement of the cerebellum with skilled movement is illustrated here in an experiment in which Chris Miall and his colleagues from Oxford used fMRI to study brain activation patterns that occur while a subject reaches for a target (Fig. 14). A specific function of the cerebellum appears to be to calculate the position of the hand in space relative to where it is intended to go, making a "forward model" that informs cortical motor areas developing the specific sequence of movements needed. A critical aspect of this forward model for which the cerebellum and its connections are specialized is in tracking the timing of movements. These functions give the cerebellum an important role in motor learning, as well as in performing motor tasks.

Macbeth describes how his eyes are fixed on the apparent reality of the blade as he complains that "Mine eyes are made the fools o'th' other senses." An important part of brain activity during a targeted movement is in the way in which the cerebellum helps integrate feedback from touch and position sense in the hand and limb with visual information to ensure that the hand reaches the target accurately. Macbeth's consternation thus is increased by the disparity between the input from his eyes and the lack of an appropriate sensory response from his hands, which are unable to clutch the "fatal vision" suspended in the empty air before him. We may experience a similar curious unreality as we try to grasp an object in a hologram or the strange virtual world of a computer game.

Like other parts of the brain, the cerebellum has a so-called somatotopic organization. That is to say, there is a direct correspondence between the precise region of the cerebellum that becomes active and the part of the body that is moved. As in other areas of the brain (most notably the primary motor and sensory cortices), the pattern of functional changes in the cerebellum defines a crude body shape (a "homunculus," or "little man"). Thus, when moving his right hand, an actor playing Macbeth would activate the right cerebellar hemisphere and the left primary motor cortex, as shown in Figure 14.

LEARNING AND GROWING

A comedy of wit and wordplay, Shakespeare's Love's Labour's Lost *begins with men making an absurd vow to abstain from the company of women for three years. King Ferdinand of Navarre joins with three of his lords—Berowne, Dumain, and Longaville—in swearing to pursue rigorous studies rather than romantic involvement. With the arrival of a French princess and her ladies, Shakespeare shows how such foolish vows, made to be broken, lead to the men's comic undoing. In the scene that follows, Berowne announces his discovery that the direct experience of love is the source of true learning.*

Berowne: Have at you then, affection's men-at-arms:
Consider what you first did swear unto,
To fast, to study, and to see no woman;
Flat treason 'gainst the kingly state of youth.
Say, can you fast? your stomachs are too young,
And abstinence engenders maladies.
And where that you have vow'd to study, lords,
In that each of you have forsworn his book,
Can you still dream and pore and thereon look?
For when would you, my lord, or you, or you,
Have found the ground of study's excellence
Without the beauty of a woman's face? . . .

O! we have made a vow to study, lords,
And in that vow we have forsworn our books:
For when would you, my liege, or you, or you,
In leaden contemplation have found out
Such fiery numbers as the prompting eyes
Of beauty's tutors have enrich'd you with?
Other slow arts entirely keep the brain,
And therefore, finding barren practisers,
Scarce show a harvest of their heavy toil;

Loves' Labour's Lost
Clockwise from upper left:
Jason Patrick Bowcutt, Alene Dawson, Libby Christophersen, Michael Medico, Enid Graham, Dallas Roberts, Melissa Bowen and Sean Pratt.

But love, first learned in a lady's eyes,
Lives not alone immured in the brain,
But, with the motion of all elements,
Courses as swift as thought in every power,
And gives to every power a double power,
Above their functions and their offices.
It adds a precious seeing to the eye;
A lover's eyes will gaze an eagle blind;
A lover's ear will hear the lowest sound,
When the suspicious head of theft is stopp'd:
Love's feeling is more soft and sensible
Than are the tender horns of cockled snails:
Love's tongue proves dainty Bacchus gross in taste.
For valour, is not Love a Hercules,
Still climbing trees in the Hesperides?
Subtle as Sphinx; as sweet and musical
As bright Apollo's lute, strung with his hair;
And when Love speaks, the voice of all the gods
Make heaven drowsy with the harmony.
Never durst poet touch a pen to write
Until his ink were temper'd with Love's sighs;
O! then his lines would ravish savage ears,
And plant in tyrants mild humility.
From women's eyes this doctrine I derive:
They sparkle still the right Promethean fire;
They are the books, the arts, the academes,
That show, contain, and nourish all the world;
Else none at all in aught proves excellent.
Then fools you were these women to forswear,
Or, keeping what is sworn, you will prove fools.
For wisdom's sake, a word that all men love,
Or for Love's sake, a word that loves all men,
Or for men's sake, the authors of these women,
Or women's sake, by whom we men are men,
Let us once lose our oaths to find ourselves,
Or else we lose ourselves to keep our oaths.

Love's Labour's Lost, 4.3

Shakespeare undoubtedly spent many schoolboy hours committing large tracts of classical texts to memory, but he was not a scholarly drudge! He was a practical playwright, picking up information from all of the sources available to him. One feels that Shakespeare—like his character Berowne—might have thought that knowledge derived from "book learning" is incomplete without a broader experience, and "learning by doing."

Learning and creating new memories are extraordinary examples of a form of brain "plasticity" that functions even through adulthood. With learning, even the mature brain continues to change over time. Understanding how this occurs has been a major focus for research employing many modern scientific tools, including functional brain imaging.

This passage celebrates the learning (at least with respect to the subject of love) that can come from the study of a woman's face, as well as that from great books. Berowne highlights an important issue here—that there are distinct forms of learning and memory. Perhaps he is referring to this when he says, "For when would you, my lord, or you, or you,/Have found the ground of study's excellence/Without the beauty of a woman's face?" suggesting that facts about the world around us may be more difficult to remember than things that happen to us or facts related to personal experiences.

More specifically, Shakespeare calls to mind the different forms of memory underlying learning, touching on distinctions recognized by modern brain science. What scientists now call declarative memory systems are used for knowledge that is not part of everyday experience but is gained in what Berowne describes as "leaden contemplation." For example, recalling that porpoises are mammals and breathe air, or that the Second World War ended in 1945, makes use of declarative memory. In contrast, "episodic" memory refers to a form of memory that is based on personal experience, like that of love "first learned in a lady's eyes." Remembering that a neighbor's dog always barks as you pass the house and recalling that you have always found Camembert too strong are examples of episodic memories.

For when
would you,
my lord, or you,
or you,
Have found
the ground
of study's
excellence
Without the
beauty of a
woman's face? . . .

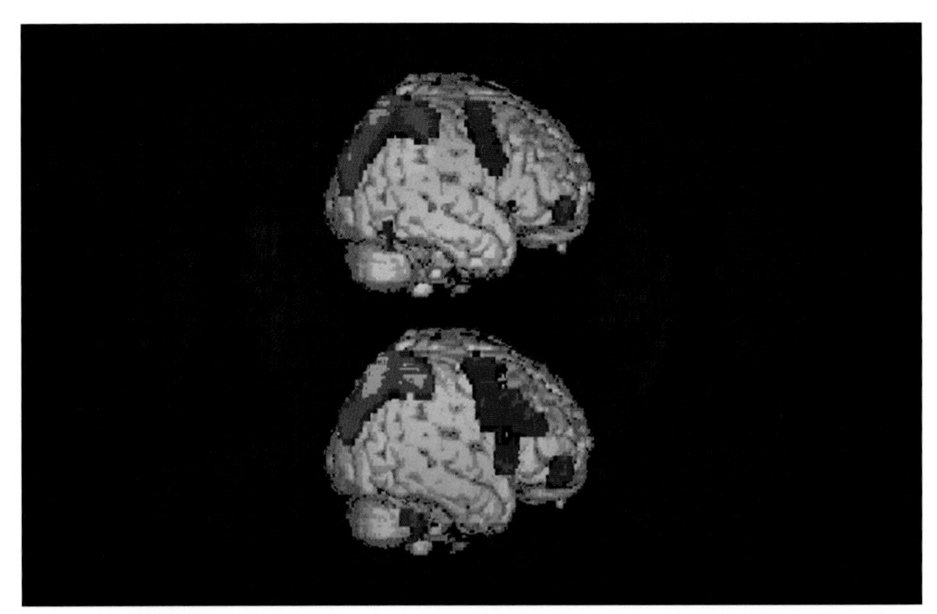

Figure 15. *The concept of "working memory" includes those processes involved in the short-term recall of information. When we look up a telephone number and keep it briefly in our minds before dialing, for example, we use working memory. Experiments like the one shown here, which used PET scanning to study the pattern of brain activation during a simple working memory task, suggest that the regions in the lower frontal lobe may be more important for retrieving information, while those higher in the frontal lobe are called into play when monitoring and manipulation of this information is necessary.*

The observation made many years ago that one of these forms of memory can be lost without damage to the other suggested that different systems in the brain must be responsible for each. In *The Body in Question,* the British physician and stage director Jonathan Miller gives a dramatic illustration of the independence of the different types of memory in the tragic story of an English musician who suffered from a terrible brain infection (or encephalitis) caused by a form of the common herpes simplex virus.

This virus attacks the brain particularly in areas responsible for functions associated with laying down new memories.

Because he became confused due to illness and wandered away from his home, this patient was not treated until well after the illness had destroyed large sections of the lower middle part of the brain (in the temporal lobes). The damage left him unable to acquire any new memories about his experiences (*episodic* memory), although he retained old memories and his memories for previously learned skills. He could not recall from one moment to the next whether he had met a visitor or even when he had awakened that morning, but paradoxically, he retained the high level of musical knowledge (stored in *declarative* memory) necessary to conduct the orchestra that he had led before his illness.

Recent brain imaging studies are beginning to shed light on how this might happen by mapping areas of the brain involved in the apparently different forms of memory and learning. The image shown in Figure 15 was obtained from a normal subject who had been asked to remember a simple picture for a short time. We use this type of short-term memory or "working memory," in most tasks that we perform through the day, whether the task is keeping a phone number in mind between looking it up and dialing, or reaching out for a towel at the sink when our eyes are shut to keep water and soap out. This image highlights the way in which these types of tasks activate the frontal lobes of the brain. These areas in the frontal lobes have a rich network of connections with the temporal lobes, which play a prominent role in longer-term declarative and episodic memory systems. Our memory for the special arrangement of objects in a room seems to be processed particularly by areas of the brain known as the hippocampus and parahippocampus. Encephalitis destroyed these areas on both sides of the brain of the patient Jonathan Miller described.

Although they are distinct, declarative and episodic memories can reinforce each other. Berowne suggests this as he reminds his friends that experience can augment the lessons of books. This potential for interaction between memory systems was embraced by classical Chinese scholars, who had to remember enormous amounts of complex and abstruse information to pass their civil service examinations, an important route to social advancement. In the seventeenth century, Portuguese Jesuit missionaries taught some of these scholars to build a so-called memory palace in their minds as a way to help them remember the large amounts of material. With this method, the scholars associated ideas that needed to be remembered with objects or places in a room of their imaginary palace. In each room individual "objects" were imagined that corresponded to specific things to be remembered. For example, to remember a sequence of words such as *car, rain,* and *garden,* one might imagine that the foyer of a memory palace has a hook for keys (car) on the wall underneath a broad-brimmed hat (rain), next to a reproduction of a Monet painting (garden). With practice (and a much longer list), this technique can be used today by those who wish to impress others with their memory power!

JUDGMENT AND CONTROL

Decision and Action

In the first of Shakespeare's two Henry IV *plays, the young Prince Hal, the son of Henry IV, and his ruffian companions epitomize irresponsibility and poor judgment. Having usurped the crown, Henry IV worries about the future of the throne, given Hal's dissolute behavior and apparent lack of interest in affairs of state. The prince prefers to spend his time "partying" rather than learning the role of a king. However, even early in the play, Shakespeare hints at the greatness that is to come. Mischievously, when Hal's friends plot a highway robbery, he arranges a comic robbery of the robbers. After the wild confusion quiets and he is left to reflect alone, Hal speaks his thoughts and reveals the noble spirit that will eventually serve him well as King Henry V. Here Shakespeare dramatically demonstrates the human capacity for developing good judgment and self-control—at any age.*

Prince Hal: I know you all, and will awhile uphold
The unyok'd humour of your idleness.
Yet herein will I imitate the sun,
Who doth permit the base contagious clouds
To smother up his beauty from the world,
That, when he please again to be himself,
Being wanted he may be more wonder'd at
By breaking through the foul and ugly mists
Of vapours that did seem to strangle him.
If all the year were playing holidays,
To sport would be as tedious as to work;
But when they seldom come, they wish'd-for come,
And nothing pleaseth but rare accidents:
So when this loose behaviour I throw off,
And pay the debt I never promised,
By how much better than my word I am,
By so much shall I falsify men's hopes;
And like bright metal on a sullen ground,

Henry IV, Part I. Derek Smith as Prince Hal.

My reformation, glitt'ring o'er my fault,
Shall show more goodly, and attract more eyes
Than that which hath no foil to set it off.
I'll so offend, to make offence a skill,
Redeeming time when men think least I will.

Henry IV, Part I, 1.2

Following the part of the play from which this quotation is drawn, Prince Hal changes from a careless youth interested only in drink and pleasure into a wise and powerful king. He leaves behind his dissolute companions (including the memorable Sir John Falstaff) and surrounds himself with mature advisers as he assumes the responsibilities of the crown, altering himself from one who thoughtlessly responds to every moment's whim into a long-term planner and a careful strategist. He makes stern moral judgments, not least of which is to condemn the "foul and ugly" tendencies of his former companions.

Dramatic tension in the play comes in part from the wonder of both the audience and the play's other characters that a grown man could so dramatically transform his character. In this soliloquy, Shakespeare challenges our common experience (captured in popular sayings such as "A leopard cannot change his spots," or "You can't teach an old dog new tricks") that it is not possible for an adult to change the brain and mind substantially.

Examples like Prince Hal's dramatic transformation for the better are known, but they seem sadly to most people to be all too rare an experience. More frequently, we learn of people who are transformed by disease or addiction from apparently levelheaded and responsible individuals into persons unable to control their cravings and appetites.

A well-known example of this is the story of Phineas Gage. In 1848, Phineas Gage was a supervisor working on railway construction in Vermont. He was renowned for his good character and reliability. One day he was preparing a dynamite charge to explode rock for construction of the line. When Gage was tamping down the explosive charge with a large metal rod, a

I'll so offend,
to make
offence a skill,
Redeeming
time when
men think
least I will.

spark flew from the metal scraping rock and ignited the dynamite. The subsequent explosion sent the three-foot metal tamping iron through his skull and the front of his brain. Remarkably, he survived.

Phineas Gage lived for twelve years after the accident, but as a changed man. He developed epilepsy as a consequence of

Figure 16. *This striking image was composed from a computer reconstruction of CT scans of the skull of Phineas Gage, who experienced a dramatic personality change after a large spike was accidentally sent through the middle part of the front of his brain. The CT work was done recently, despite Gage's death almost a hundred years ago. In the laboratory of Hannah and Antonio Damasio, at the University of Iowa, the reconstructed image was used to determine the precise course of the projectile, indicated in bright red. This allowed them to identify clearly that the area of the brain injured was the area we now associate with forward planning and an ability to develop an understanding of the consequences of actions.*

the massive scar in his brain, but the alteration of his personality was even more dramatic. Formerly equable, he became irritable and quick to anger, an alcoholic who had cravings too base for commentators of the time to dwell on. He was no longer able to hold down a job for any length of time because, despite knowing better, he could not be depended on to act responsibly.

Not along ago, the brain scientists Hannah and Antonio Damasio and their colleagues at the University of Iowa revisited this mystery. Taking Phineas Gage's skull from its repository in a museum in Cambridge, Massachusetts, they performed CT scans and then electronically reconstructed the surface of the skull in order to define the precise path taken by the tamping iron (Fig. 16). By relating brain to skull structure, they were able to demonstrate that the iron must have damaged the orbitofrontal and prefrontal cortices.

The orbitofrontal cortex in the lower part of the front of the brain, near where the tamping iron entered Phineas Gage's brain, is part of an internal reward system that helps us put off the need for immediate gratification. Areas just above this in the prefrontal cortex, also damaged by Gage's injury, are important for focusing attention on tasks and shifting attention between different parts of tasks to complete them successfully. Together these areas seem to be responsible for the functioning of what we call a moral sense. With such extensive damage to these areas, there should be no wonder that the character of Phineas Gage was altered.

Prince Hal's soliloquy shows that he is deliberately hiding his true feelings from others. This involves continued close monitoring of what he allows himself to say and how he allows himself to act in the company of his friends. Such activity must also occur in the prefrontal cortex. For this reason, some have suggested that brain imaging studies that test the function of this area might even be used some day as a modern and more sophisticated form of lie detector.

MOTIVATION AND MORALITY

What motivates us to act (or not to act) is a central concern of Shakespeare's drama. Among the earliest of Shakespeare's history plays, Richard III *is also the only one of his plays to begin with a soliloquy by its title character. As the Duke of Gloucester and brother to King Edward IV, Richard makes his motivation all too clear in this introductory speech. Punning on "sun" and "son" in the opening sentence, he goes on to boast of his immoral scheme to usurp the crown, excusing his evil intentions with his physical deformity. He has set his plan in motion by promoting a prophecy that someone with the initial G will murder Edward's heirs. As a result, the king orders the arrest of George, the Duke of Clarence, brother to both the king and Richard, and unwittingly aids his evil brother's rise to become Richard III.*

Richard III: Now is the winter of our discontent
Made glorious summer by this son of York;
And all the clouds that lour'd upon our House
In the deep bosom of the ocean buried.
Now are our brows bound with victorious wreaths,
Our bruised arms hung up for monuments,
Our stern alarums chang'd to merry meetings,
Our dreadful marches to delightful measures.
Grim-visag'd War hath smooth'd his wrinkled front:
And now, instead of mounting barded steeds
To fright the souls of fearful adversaries,
He capers nimbly in a lady's chamber,
To the lascivious pleasing of a lute.
But I, that am not shap'd for sportive tricks
Nor made to court an amorous looking-glass;
I, that am rudely stamp'd, and want love's majesty
To strut before a wanton ambling nymph:

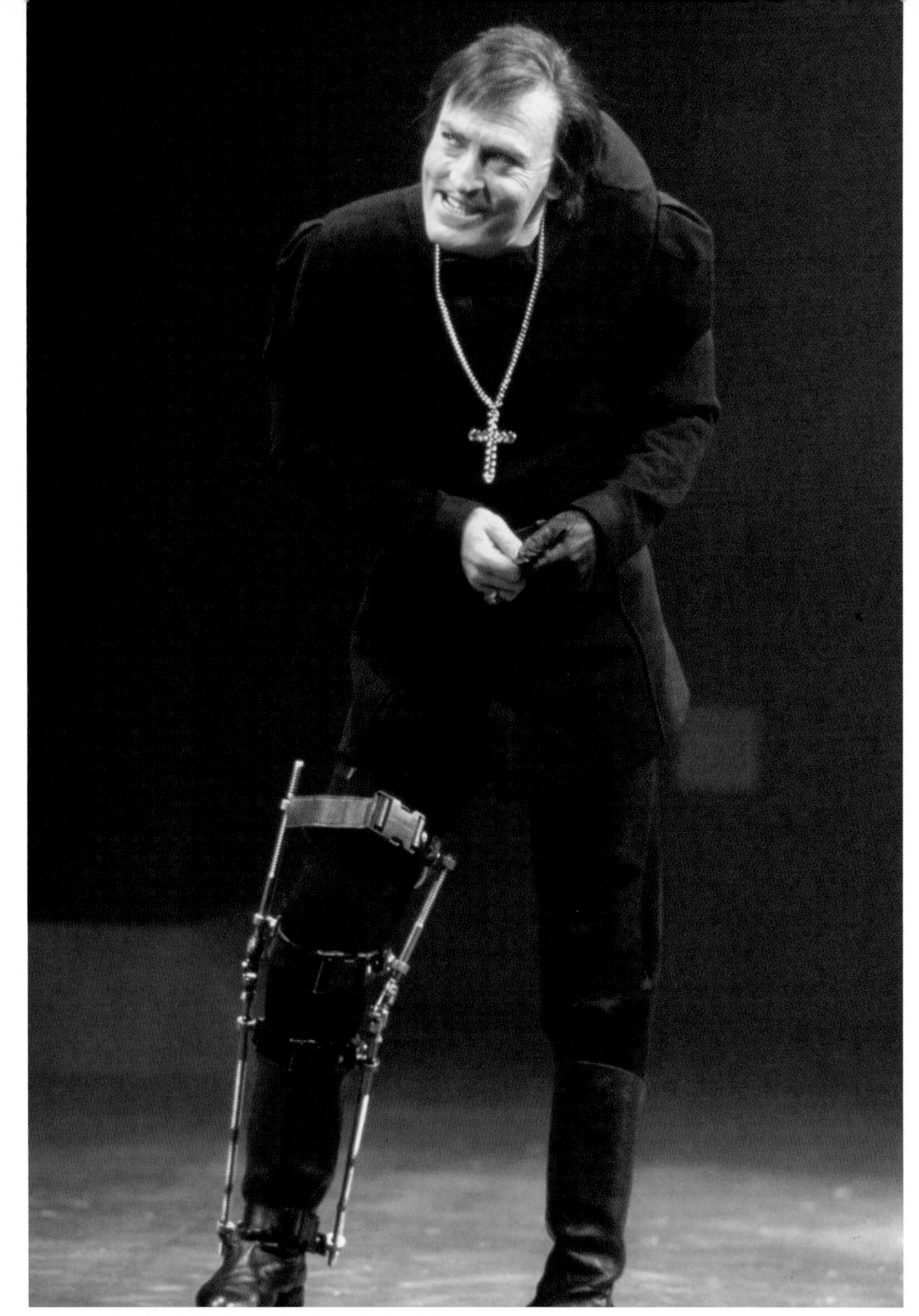

Richard III. Stacy Keach as Richard III.

And therefore, since I cannot prove a lover To entertain these fair well-spoken days, I am determined to prove a villain, And hate the idle pleasures of these days.

I, that am curtail'd of this fair proportion,
Cheated of feature by dissembling Nature,
Deform'd, unfinish'd, sent before my time
Into this breathing world scarce half made up—
And that so lamely and unfashionable
That dogs bark at me, as I halt by them—
Why, I, in this weak piping time of peace,
Have no delight to pass away the time,
Unless to spy my shadow in the sun,
And descant on mine own deformity.
And therefore, since I cannot prove a lover
To entertain these fair well-spoken days,
I am determined to prove a villain,
And hate the idle pleasures of these days.
Plots have I laid, inductions dangerous,
By drunken prophecies, libels, and dreams,
To set my brother Clarence and the King
In deadly hate, the one against the other:
And if King Edward be as true and just
As I am subtle, false, and treacherous,
This day should Clarence closely be mew'd up
About a prophecy, which says that 'G'
Of Edward's heirs the murderer shall be—
Dive, thoughts, down to my soul: here Clarence comes.

Richard III, 1.1

Richard III is particularly chilling as he describes his joint deformities of body and mind: "I, that am curtail'd of this fair proportion,/Cheated of feature by dissembling Nature, . . . I am determined to prove a villain." The horror of this passage is that a man could be so evil as to desire to harm others without specific cause. There is no motivation for revenge or power behind Richard's plan—he believes that he is "subtle, false, and treacherous" because to be so is quite simply in his nature.

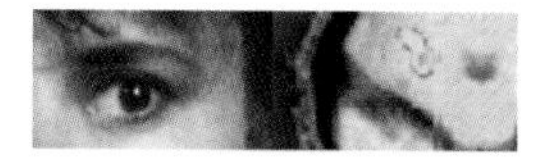

Understanding the basis of moral action has been a challenge for Western literature since its beginnings in the great epics of the *Iliad* and the *Odyssey*. It is now becoming an important issue for modern brain science. We cannot help being fascinated by the enigma of Richard, who so clearly understands our moral universe and yet rejects its tenets. There is still no science that can explain the evil of Shakespeare's Richard, but brain scientists are beginning to define structures in the brain that are essential for aspects of moral and responsible behavior.

A critical aspect of moral behavior is the ability to forgo smaller, short-term gains in order to realize more distant but greater rewards. For example, a student learns to forgo the pleasures of a night out before examinations in order to achieve higher marks and the praise and new opportunities that arise from them. Patients who have suffered severe damage to the middle part of the frontal lobe such as Phineas Gage (see "Judgment and Control," page 85) have a childlike impulsiveness that prevents them from acting "responsibly" in this way. Nonetheless, patients who damage their brains as adults, like Gage, are at least able to appreciate the concepts behind this strategy.

Recent studies by Hannah and Antonio Damasio and their group at the University of Iowa suggest that this may not be the case if damage to the same areas of the brain occurs early in development. They studied two young adults who had sustained damage to the front part of the brain (the orbitofrontal and medial frontal cortices) before 16 months of age, as shown in Figure 17 (overleaf). Both had a history of severe family and social problems arising from irresponsibility, an inability to follow orders, and lack of guilt or remorse for misdeeds. However, not only were these subjects impaired in their moral behavior, like those who suffer similar lesions as adults, but they also showed defects in their ability to perform social and moral reasoning in the abstract. With early damage to this area of the brain, they simply were unable to appreciate the rules. These cases were clearly different from that of Phineas Gage (Fig. 16, page 86), who retained an ability to distinguish between right and wrong.

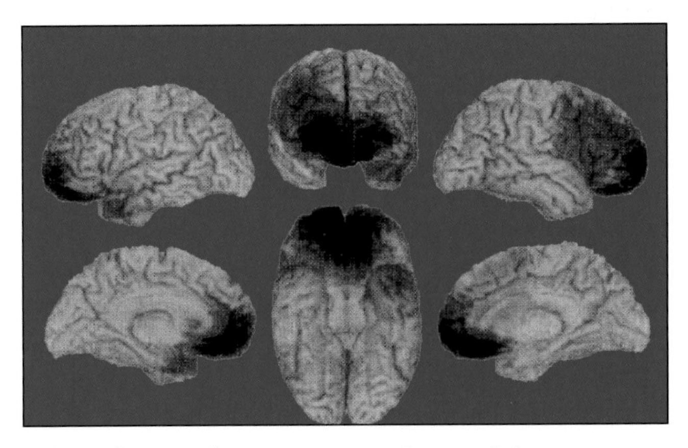

Figure 17. *Extending their work in trying to understand the reasons for the changes in personality in Phineas Gage (see Fig. 16), the Damasio group went on to assess the consequences of damage to the same areas of the brain at a very young age. They studied two patients who had suffered from injuries to the middle front part of the brain in the first years of life. The stories of these individuals are sad because they were unable to integrate properly into society, and had histories of irresponsible and even criminal behavior. Unlike Phineas Gage, who was thought to be able to appreciate that some actions would likely lead to good outcomes and others to bad (despite an inability to reliably choose those with a better outcome), these subjects with early injury to similar areas of the brain (shown by the darkening on the brain images above) seem unable to appreciate the concepts of right and wrong. This suggests that development of the mental tools necessary for moral behavior may demand that these structures be intact in the brain during a critical formative period.*

The conclusion that damage to the front of the brain may keep people from being able to understand why their actions are right or wrong has far-reaching social implications. Figure 18 includes PET scans made by the California scientist Adrian Raine of the brains of murderers and of a healthy, law-abiding person. The images show decreased metabolism in the brain of one of the murderers, particularly in the frontal regions. This suggests that the individual had decreased brain functioning in just the areas that, when damaged, were shown by the Damasios to be associated with morally deficient behavior. Some people now wonder

whether information regarding the functioning of this part of the brain is important to consider in judging the responsibility of an individual for his crime (rather as Richard III suggests to the audience when he claims a defense for his evil based on his deformities). Does a person with a poorly functioning frontal lobe have diminished responsibility for a crime because of a biological

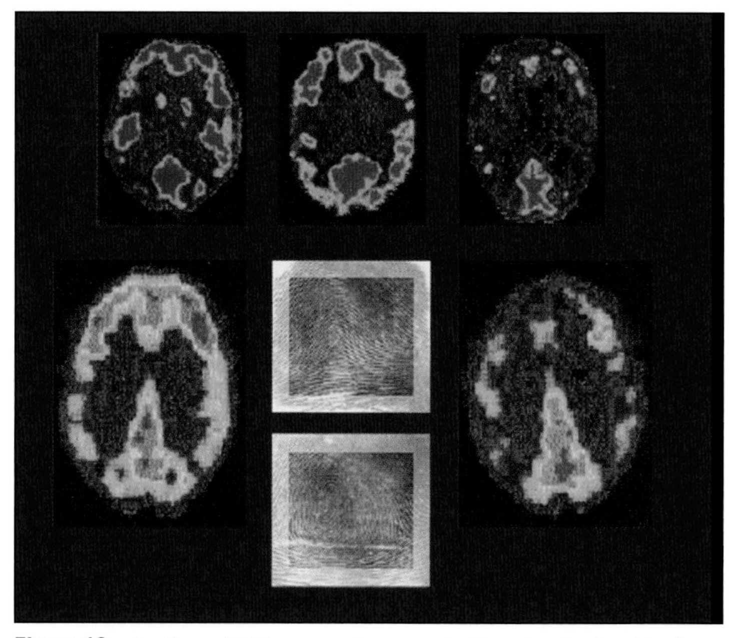

Figure 18. *In these PET scans the metabolism in the brain of a normal volunteer (upper left) is contrasted with similar images from brains of murderers. The middle upper image is an example of a brain scan from a murderer who came from a socially deprived background. It shows a normal level of metabolism. The upper right scan is from a murderer from a non-deprived background that shows severe hypometabolism in the front part of the brain. Some scientists believe that the abnormal scan is evidence that the sense of right and wrong may be impaired because of an underlying dysfunction of the brain. In contrast, criminal behavior with normal brain function may be more a consequence of sociopathic personality developed because of environmental influences. However, interpretation of such data remains controversial. A similar example of a murderer in which the front part of the brain again shows impaired metabolism is shown on the lower right-hand side (compare the normal image at the lower left).*

impairment in the ability to distinguish right from wrong?

Most neuroscientists would probably suggest that we are a very long way from being able to predict behaviors or capacities solely from changes seen in brain imaging. But if a defendant faces capital punishment, should one risk overinterpretation of the scientific evidence?

Another question lies behind Richard's defense of his actions based on his having been "Cheated of feature by dissembling Nature." Was he evil because he was as deformed in mind as he was deformed in body, or did the way in which people treated him because he was imperfect cause him to be evil? The relationship between nature and nurture remains a lively topic of debate. To address a relevant aspect of this problem, Raine and his colleagues have begun to test the relationship between environmental factors that might predispose to violence (such as the raising of a child in a "bad" home) and the patterns of functional abnormalities that are seen in the brain. The PET images in Figure 18 suggest that impaired frontal brain activity ("nature") might make a greater contribution to the risk of criminality in a child from a "good" home than one from a "bad" home (where "nurture" might be argued to have a potentially larger role).

Shakespeare clearly believes Richard to be an evil man and demands that we allow ourselves to make a moral judgment of him. In real life, however, the situation is generally more ambiguous. To what extent can we accept that the human will is limited by the integrity of the brain? To the extent that we do, is there an important role for functional brain imaging in administering our laws justly, as the work of Adrian Raine may suggest? The answers to these questions are neither obvious nor simple. They must be addressed jointly by scientists and other citizens as cognitive brain science develops.

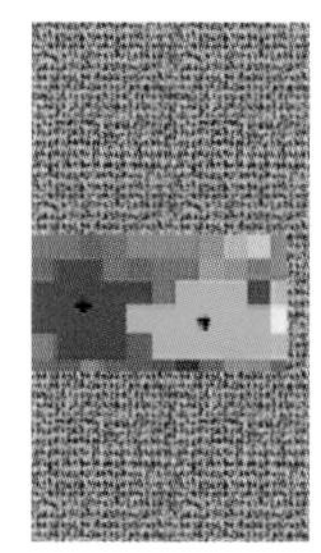

4. Language and Numbers

A MUSE OF FIRE

The richness of the mental images conjured by Shakespeare's words makes a major contribution to the poetry of his plays. Word pictures light up the opening of Shakespeare's Henry V, *and its prologue is filled with tantalizing imagery of the play that follows. Continuing the story of Prince Hal's transformation that was begun in the earlier* Henry IV *history plays,* Henry V *presents the patriotic picture of a warrior king leading his army to success against the French. In this introductory speech, the phrase "this wooden O" refers to the type of circular theater popular in Elizabethan England. This design, which thrusts the actors into the audience, can now be experienced just as Shakespeare knew it, in London's faithful reconstruction of the Globe Theatre.*

Chorus: O for a muse of fire, that would ascend
The brightest heaven of invention,
A kingdom for a stage, princes to act,
And monarchs to behold the swelling scene!
Then should the warlike Harry, like himself,

Assume the port of Mars, and at his heels,
Leash'ed in like hounds, should famine, sword and fire
Crouch for employment. But pardon, gentles all,
The flat unraised spirits that hath dared
On this unworthy scaffold to bring forth
So great an object. Can this cockpit hold
The vasty fields of France? Or may we cram
Within this wooden O the very casques
That did affright the air at Agincourt?
O pardon, since a crooked figure may
Attest in little place a million,
And let us, ciphers to this great account,
On your imaginary forces work.
Suppose within the girdle of these walls
Are now confined two mighty monarchies,
Whose high upreared and abutting fronts
The perilous narrow ocean parts asunder.
Piece out our imperfections with your thoughts.
Into a thousand parts divide one man
And make imaginary puissance.
Think, when we talk of horses, that you see them
Printing their proud hoofs i'th' receiving earth.
For 'tis your thoughts that now must deck our kings,
Carry them here and there, jumping o'er times,
Turning th'accomplishment of many years
Into an hour-glass: for the which supply,
Admit me Chorus to this history,
Who prologue-like your humble patience pray,
Gently to hear, kindly to judge our play.

Henry V, Prologue

Think, when we talk of horses, that you see them Printing their proud hoofs i'th' receiving earth.

For the actors, Shakespeare himself is the muse of fire whose inspiration the Chorus attempts to awaken. It is his language that gives the actors power to move the audience and make the stage into a "kingdom" broad with the "swelling scene." In *Henry V*, the richness of Shakespeare's language, triumphantly heralded by the Chorus, brings "the vasty fields of France" to

Henry V. Wallace Acton as a Chorus member.

the stage. One of the mysteries evoked by this passage, and now beginning to be probed by brain science, is the way in which language can conjure strong images in the mind.

The first figure here illustrates a remarkable imaging study of responses of the brain to language (Fig. 19). Magnetoencephalographic (MEG) measurements of brain activity are superimposed

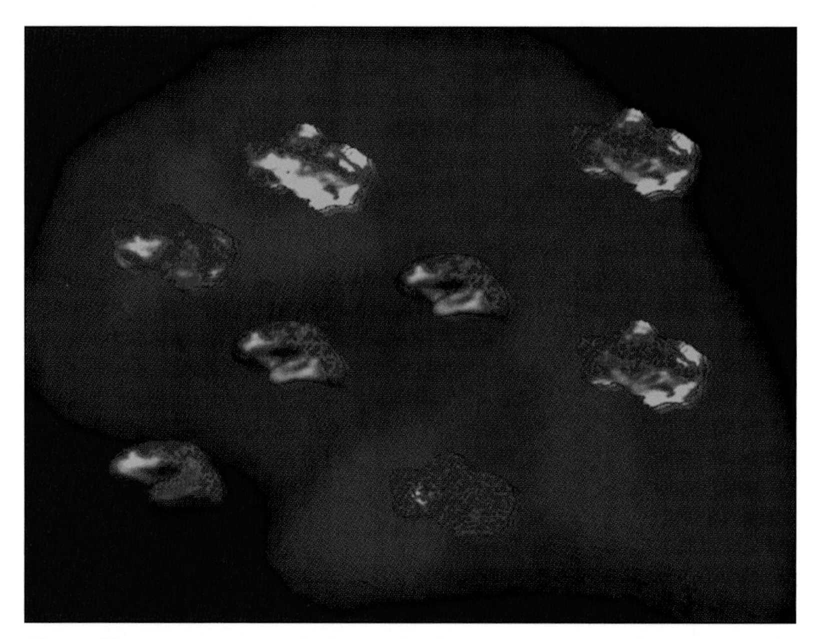

Figure 19. *Magnetoencephalography (MEG) is a powerful new tool for imaging activity of the brain. Sensitive detectors (supercooled inside to more than 300 degrees below zero Fahrenheit) placed next to the skull measure the very tiny magnetic fields associated with electrical signaling activity in the brain. The pattern of these signals can be used to develop images of brain activity such as those shown here. Combined use of MEG and functional magnetic resonance imaging (fMRI) allows the signals to be localized both more accurately (that is to say, better reflecting the "truth") and more precisely (or with lower uncertainty). The images shown here display brain activity during a semantic (word meaning) task (see also Fig. 21). The areas of high activity are shown in orange-yellow on either inflated (middle images) or flattened brains, which show the relationships between different activated regions more clearly than does the highly convoluted structure of the normal brain.*

on MRI images of the brain that have been processed to remove the complex surface convolutions either by being blown up like a balloon or being flattened like a road map. The individual images show brain activity when a subject read either entirely new or repeated words and made a decision about their meaning (see also "A Subtle Voice," page 101, and Fig. 21). With both new and repeated words, reading activated the primary sensory and motor areas of the brain, as well as the visual cortex. Part of this activation is due simply to the brain processes involved in reading—for example, coordination of eye movements with serial perception of the individual letters and word forms. However, processes involved in assigning meaning to the words also make a substantial contribution.

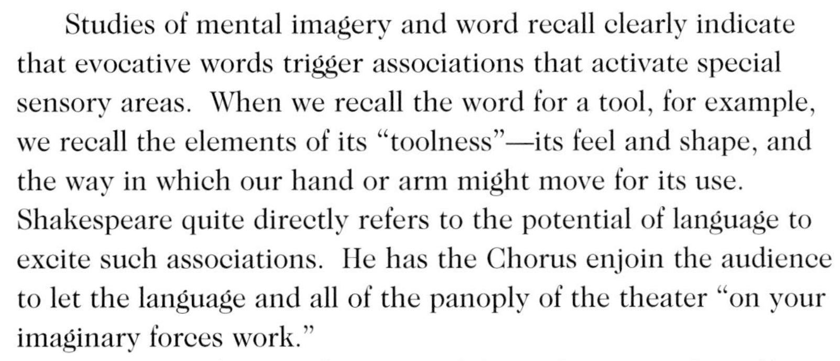

Studies of mental imagery and word recall clearly indicate that evocative words trigger associations that activate special sensory areas. When we recall the word for a tool, for example, we recall the elements of its "toolness"—its feel and shape, and the way in which our hand or arm might move for its use. Shakespeare quite directly refers to the potential of language to excite such associations. He has the Chorus enjoin the audience to let the language and all of the panoply of the theater "on your imaginary forces work."

The second figure is from one of the early PET studies of brain activation, in which subjects were presented with either written or spoken words. The image on the left in Figure 20 (overleaf) shows brain activation associated with reading, which is understandably prominent in the primary visual cortex in the back of the brain.

Activation of areas of the brain responsible for interpreting words that are heard is shown in the image on the right. Here the most prominent activation occurs on the side of the brain, where sounds are processed in the primary auditory cortex. The differences in the pattern of brain activation between reading and hearing words highlight the differences in the perceptual processes. Nevertheless, there are also regions of activation along the left side of the brain in areas that are associated with the interpretation of meaning of both written and heard language. The

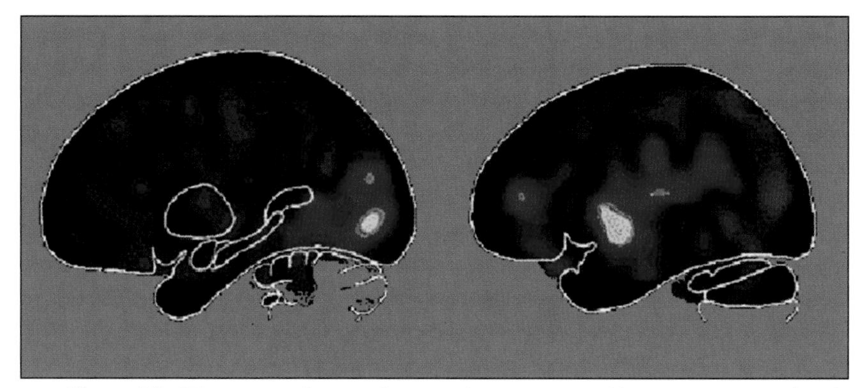

Figure 20. *This now classical pair of PET scans illustrates very clearly the different ways in which the brain works to perceive words. The image on the left side was made after the subject was asked to read words, which led to activation in the back of the brain, where the visual cortex is found. For the image on the right, the subject listened to words, which activates the auditory cortex, located along the middle of the brain. In both cases, the language system was activated, but the form of presentation of the words was different. The primary perceptual event therefore is critical in determining the initial path for the processing of language.*

activation of this final, common language system, whether the words are read or heard, helps explain how the power of Shakespeare's language comes across not only as we read his plays quietly in an armchair, but also when we are part of an audience in a theater.

Distinct areas around the upper part of the temporal lobe and the frontal lobe on left side of the brain begin to "fire" with the use of language. Recent evidence suggests that distinct pathways may be involved in identifying words as potentially meaningful and in assigning them meaning. An "imaginary puissance" (meaning strength or power) appears to come from the linking, or *binding,* of activations associated with responses to the sometimes multiple properties that give meaning to a word (see "Binding Qualities in Meaning," page 111). In this way, the findings of modern brain science would directly support the urging of the Chorus: "Think, when we talk of horses, that you see them/Printing their proud hoofs i'th' receiving earth."

A SUBTLE VOICE

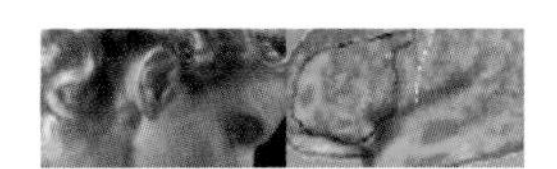

Language and Numbers

Word choice is key to the poet's art. Novel words or turns of phrase are so striking to us that their novelty must influence the way in which the brain processes language. In The Merchant of Venice, *Portia speaks one of Shakespeare's most emotionally charged passages, which includes unexpected and novel word images. Dressed as a man, Portia enters a Venetian courtroom and passes herself off as Balthazar, a young doctor of laws from Rome. In the passage chosen here, Portia eloquently reminds the court of the "quality of mercy" that "droppeth" like rain and urges Shylock to be merciful toward the merchant Antonio. It is important to know that although Shylock remains unmoved by her impassioned arguments and demands that Antonio forfeit the promised pound of flesh, Portia's subtle voice eventually works its power and turns the tables on Shylock.*

Portia: The quality of mercy is not strain'd,
It droppeth as the gentle rain from heaven
Upon the place beneath: it is twice blest.
It blesseth him that gives, and him that takes,
'Tis mightiest in the mightiest, it becomes
The throned monarch better than his crown.
His sceptre shows the force of temporal power,
The attribute to awe and majesty,
Wherein doth sit the dread and fear of kings:
But mercy is above this sceptred sway,
It is enthroned in the hearts of kings,
It is an attribute to God himself;
And earthly power doth then show likest God's
When mercy seasons justice: therefore Jew,
Though justice be thy plea, consider this,
That in the course of justice, none of us
Should see salvation: we do pray for mercy,

The Merchant of Venice. Enid Graham as Portia and Hal Holbrook as Shylock.

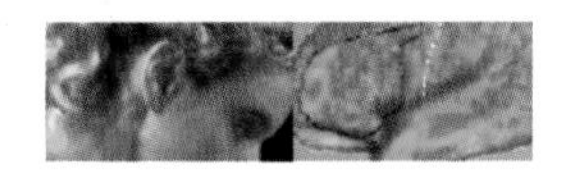

And that same prayer, doth teach us all to render
The deeds of mercy. I have spoke thus much
To mitigate the justice of thy plea,
Which if thou follow, this strict court of Venice
Must needs give sentence 'gainst the merchant there.

The Merchant of Venice, 4.1

One aspect of great poetry such as that of Shakespeare is the way in which the words resonate in our minds. As we read, "The quality of mercy is not strain'd/It droppeth as the gentle rain from heaven/Upon the place beneath," the words arrest our attention because of the beauty and uniqueness of both the metaphor and the sound. Why is it that the novelty of Shakespeare's words and their combinations adds so much to his poetry?

The quality of mercy is not strain'd, It droppeth as the gentle rain from heaven Upon the place beneath: it is twice blest.

We respond to novelty in different ways than we respond to the over-familiar, and this is as true for words as it is for other inputs to the brain. The accompanying figure illustrates this (Fig. 21; overleaf). In creating these images, brain scientists used the sensitive technique of magnetoencephalography (MEG) to study the patterns and relative timing of activations in the brain during a task that involved making a simple decision (see also Fig. 19). Recall that the MEG method detects the very tiny changes in magnetic fields produced by electrical activity in neurons. Subjects were asked to read a word that was presented on a television monitor in front of them. They then were asked to judge whether the object described was more than a foot long in any dimension. It is extraordinary how much of the brain was activated even with this simple task. Imagine how much more of the brain we must call into play as we decode the meaning of and establish the associations between the evocative words of Shakespeare!

A key aspect of this study was that half of the words presented were not expected by the viewers, while the subjects had been shown the other half previously. Similar areas of activa-

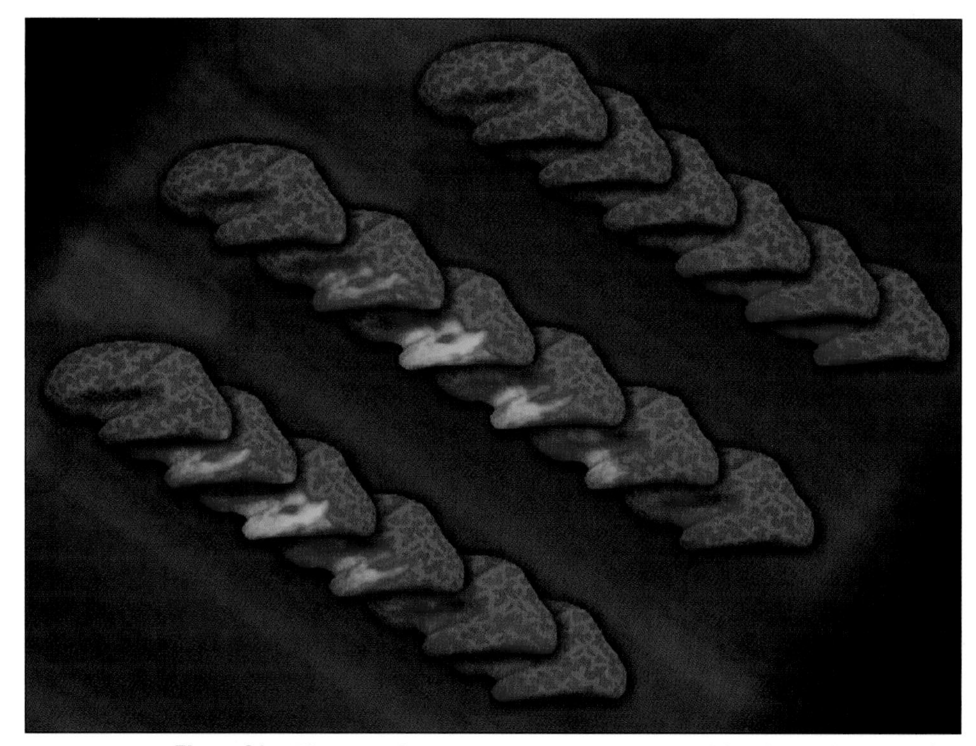

Figure 21. *These striking images were generated by the technique of magnetoencephalography (MEG), which monitors patterns of brain activity from the small changes in magnetic fields that they produce at the surface of the skull. Each successive image represents the pattern at a slightly later time after presentation of the stimulus. In this case, the subjects were asked to repeatedly read words and decide whether or not the object or animal represented by the word was larger than a foot long in any dimension. Some of the words were always new, while others were presented again and again. Similar patterns of brain activation were elicited by both novel and repeated words, but activations for the repeated words were less long-lived than those for the novel words.*

tion in language centers of the brain were found with both. However, a clear difference between responses to novel and to repeated words was found. Activations for the repeated words lasted for a much shorter time; brain activations provoked by novel words were larger and much more sustained.

Shakespeare intuitively made use of this concept to make particular concepts or sounds reverberate through a passage. Consider, for example, the sound of the phrase "mercy is above this sceptred sway," with its gentle "s" repetition and the complex use of the word *sway,* which refers simultaneously to an earthly monarch's power and to a shifting, side-to-side movement like that of a tall building in the wind. A particularly interesting feature of the MEG results is that differences between activations for repeated and for novel words were found in all parts of the language network. Even the areas involved in the earliest stages of perceptual processing showed a difference between the two. This suggests that with novel presentations of words, higher brain centers must allow us to focus on words at a very early stage in their processing by the brain. It is a true "arresting" of our attention at the most primary level.

This is just one example of a series of observations showing that when our attention is aroused, activities in many or all of the areas of the brain involved in the process are altered.

When we pay close attention to detecting a light touch, for example, brain activation is increased in most areas of the brain that respond to touch. Part of the response to novelty therefore could be a consequence of the increased attention that tends to be paid to novel stimuli. When you next listen to a play, consider how important novelty of presentation is for the actors as well. The responses of the audience to the playwright's words are determined by the ways in which emphasis is provided by voice, inflection, and actions, as well as by the words themselves.

WHAT IS A WORD?

How does the brain associate meaning with a word? The complexity of the problem of semantics is illustrated by Sir John Falstaff's bitter soliloquy in Henry IV, Part I. *Falstaff (the lazy and dissolute knight who had befriended the youthful Prince Hal) tries to lead a ragtag company of foot soldiers to war against northern rebels. Before the battle, he fearfully seeks the protection of the prince, who brusquely rebukes him and tells him to fend for himself. Left alone and feeling hurt, Falstaff speaks this cynical but memorable monologue (what he calls "my catechism") on the meaning of the word* honor, *which he decides is nothing more than a scutcheon (a heraldic device that for him has no real significance).*

Falstaff: 'Tis not due yet, I would be loath to pay him before his day—what need I be so forward with him that calls not on me? Well, 'tis no matter, honour pricks me on. Yea, but how if honour prick me off when I come on, how then? Can honour set to a leg? No. Or an arm? No. Or take away the grief of a wound? No. Honour hath no skill in surgery then? No. What is honour? A word. What is in that word honour? What is that honour? Air. A trim reckoning! Who hath it? He that died a-Wednesday. Doth he feel it? No. Doth he hear it? No. 'Tis insensible, then? Yea, to the dead. But will it not live with the living? No. Why? Detraction will not suffer it. Therefore I'll none of it. Honour is a mere scutcheon—and so ends my catechism.

Henry IV, Part I, 5.1

Shakespeare certainly did not want us to miss the irony of Falstaff's choice of the word *honor* for his musings. The word slips out in the third line of his speech. Falstaff pauses as he

Henry IV, Part I. Sam Waterston as Henry V and Stacy Keach as Falstaff.

recognizes this wonderful word *honor.* He then recalls the characteristics of honor, at least as they appear to him after his disgrace. That others might interpret honor differently contributes to the bitterness of the passage.

Reading a word, recognizing the word as having a meaning, and recalling that meaning are all very different language tasks. To give a simple example, we can read Lewis Carroll's poem "Jabberwocky" ("'Twas brillig, and the slithy toves/Did gyre and gimble in the wabe") and appreciate the appropriateness of the nonsense word forms yet not necessarily be able to recognize any meaning in them. Nonsense words can be formed because of our appreciation of patterns of the letters that are associated with words in English. On the basis of experience, we can recognize unusual words and recall that they have meaning but be unable to recall the meaning. (Being able to come up with such words on occasion may be one of the secrets of winning at *Scrabble.*) But recollection of a word's definition clearly involves a different mental strategy, for the meaning of a word carries with it memories of associated concepts, properties, and feelings.

What is honour? A word. What is in that word honour? What is that honour? Air. A trim reckoning!

Endel Tulving and his colleagues at the University of Toronto have studied the brain basis of a related phenomenon: the way in which the brain recalls associations between words. Using PET imaging, they sought to distinguish the patterns of activation in the brain when individuals recognized word pairs as having been read before in a previously studied group (Fig. 22). The processes are components of what we do as we reread and remember Shakespeare's plays.

This type of experiment has shown that word recognition and the recollection of a specific association between words share common regions of activation but are also associated with changes in different areas of the brain. Both word recognition and word recall tasks activated frontal regions including a central region known as the anterior cingulate cortex, which is associated with aspects of attention. Recognition differed from recall chiefly in that the primary perceptual areas of the brain were activated relatively more prominently. This is perhaps not

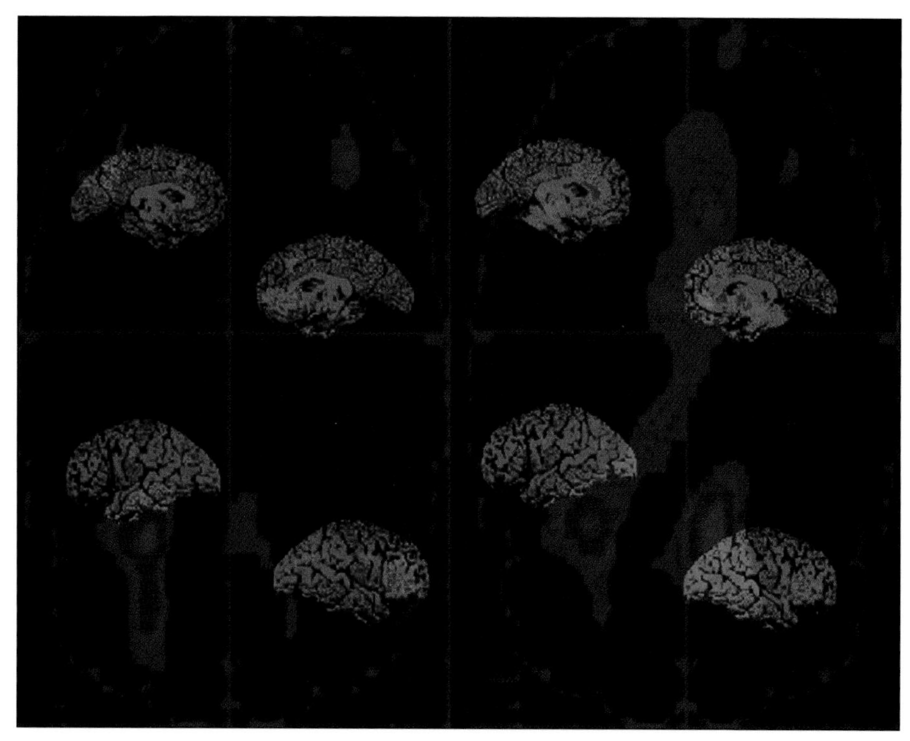

Figure 22. *These images were acquired using PET. Changes in blood flow in the brain were measured as people were asked either to recall previously studied pairs of words or to simply recognize that they had seen the words before. The act of recalling words gave rise to activation in more of the brain (shown as bright areas in the images on the right-hand side of the figure). Recognition of words activated an area in the lower part of the back of the brain associated with perception. In both cases, words are seen to activate a broad network in the brain.*

surprising, as recognition is a perceptual association task. Recalling associations between words triggers patterns of activation that are distinct from those of word recognition alone; recollection also involves regions deep in the brain. The network of interacting brain areas responsible for this word association recall is particularly large, rather as if the amount of brain involved is proportional to the relative "thoughtfulness" of the task. As might be expected if large areas of the brain are involved, clinical observations show that recalling associations between words is more susceptible to impairment from brain injury than is word recognition.

Recognition of a word and even recollection of an association between words do not need to be coupled to a common understanding of the word's meaning. Despite Falstaff's familiarity with the word *honor,* his bitter conclusion that it is no more than a "mere scutcheon" tells us that he is unwilling to give the word the value the rest of the world gives it—an attempt at self-delusion by a man whose pride has been hurt. This should not surprise us, coming from the most memorable of Shakespeare's scoundrels.

BINDING QUALITIES IN MEANING

Language and Numbers

As a poet and playwright, Shakespeare was keenly aware that we bind many associations into the meaning of a specific word or into a name. When the narrator of Sonnet 18 *asks, "Shall I compare thee to a summer's day?" the reader is moved to reflect on warm and sunny images, associations bound to the meanings of these words by common usage. In the balcony scene of* Romeo and Juliet, *Juliet triggers for us a reflection on the ways in which the brain may respond to a given word or name. Overheard by Romeo in the orchard below her, she poses her famous question: "O Romeo, Romeo, wherefore art thou Romeo?" It is easy to misread this as asking where Romeo is; in fact, Juliet asks why ("wherefore") the young man she loves must carry the name of a family that is a bitter enemy of her own. As she reasons, "That which we call a rose/By any other word would smell as sweet," Juliet presents an enduring case for words as a form of mental shorthand.*

Juliet: O Romeo, Romeo, wherefore art thou Romeo?
Deny thy father and refuse thy name.
Or if thou wilt not, be but sworn my love
And I'll no longer be a Capulet.

Romeo: Shall I hear more, or shall I speak at this?

Juliet: 'Tis but thy name that is my enemy:
Thou art thyself, though not a Montague.
What's Montague? It is nor hand nor foot
Nor arm nor face nor any other part
Belonging to a man. O be some other name.
What's in a name? That which we call a rose
By any other word would smell as sweet;
So Romeo would, were he not Romeo call'd,

Romeo and Juliet. Laura Hicks as Juliet.

Retain that dear perfection which he owes
Without that title. Romeo, doff thy name,
And for thy name, which is no part of thee,
Take all myself.

Romeo and Juliet, 2.2

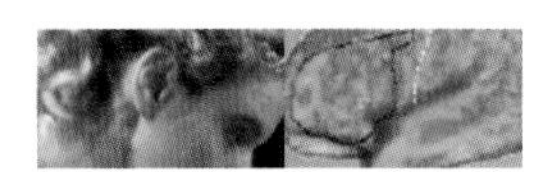

What's in a name? That which we call a rose By any other word would smell as sweet; So Romeo would, were he not Romeo call'd, Retain that dear perfection which he owes Without that title.

Throughout her young life, Juliet must have been reminded on an almost daily basis that her family's rivals, the Montagues, were all scoundrels—or worse. One can imagine her sense of confusion as she finds that this family prejudice does not correspond with her experience of the first Montague she meets, the handsome and charming Romeo. Such conflicts of experience with previously accepted beliefs are, of course, a major part of the personal growth of every adolescent. What is extraordinary about Juliet is her sophisticated approach to resolving this conflict.

Shakespeare has Juliet ponder what defines a "Montague." She rejects the idea that being a Montague defines a permanent quality of her new love, asking Romeo to "deny thy father and refuse thy name." She makes a distinction between the man and what he is called. Romeo is clearly a man of flesh and blood to Juliet, as she tenderly reviews his physical attractions in her mind. But there is no such physical reality to being a Montague, which is neither "hand nor foot/nor arm nor face nor any other part/Belonging to a man." She argues that her Romeo is a seamless composite of elements drawn from his emotional and intellectual being, as well as his physical substance. In loving her, Romeo can thus be considered independently from his family and loved in return, despite the hatred of her family for all Montagues. With these musings, Shakespeare explores a fundamental issue in understanding language: how we endow words with meaning.

For Juliet, the word *Romeo* effortlessly calls forth memories of the features and the recent tender experiences that define this man for her. In addition to the physical qualities she describes, she must also have instantly recalled the sound of his voice, his loving words, the touch of his hand, and even, perhaps, his smell. For Juliet, the combination of these qualities defines Romeo.

This passage illustrates how the binding together of associations generates meaning for a name or a word. This ability to link multiple dimensions or types of experience in a single word underlies the richness of literary creation. All writers manipulate the associations recalled by each word, and Shakespeare does this with extraordinary skill. In fact, it may be claimed that a word itself has no meanings but those that lie in the associations that the word triggers in the mind. Meaning then can be highly individualized. Thus it is that Juliet can question, "What's in a name?"

The particular immediacy of Shakespeare's poetry arises from the ability of words to call forth mental experiences similar (or perhaps even identical) to those associated with the direct experiences referred to by the words themselves. Recent research suggests that the so-called semantic (or "word meaning") memory system apparently works in part by having individual words activate mental experiences that contribute to defining the underlying concept expressed by the word. It is the binding together of these mental experiences that provides meaning.

Neuroimaging studies have shown that the recollection of words representing concrete objects activates regions in the temporal lobes that act as association areas for a variety of complex sensory processes (Fig. 23). As we have seen (see Fig. 9), a particular area of the temporal lobe is activated when a face is viewed, for example. Other areas are responsible for associating sounds with specific experiences. Sometimes activation of small areas in the brain can call remarkably specific experiences into consciousness. For example, the Canadian neurosurgeon Wilder Penfield described how electrical stimulation of a small area of the temporal lobe during surgical treatment for epilepsy made one patient hear a specific musical phrase! The parts of the temporal lobe that are activated during the recall of words therefore vary with the type of word.

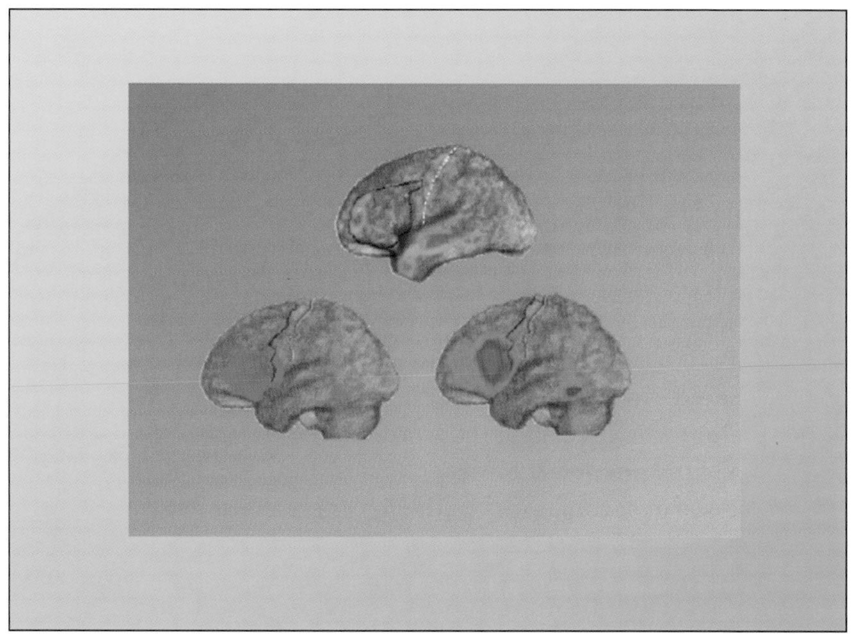

Figure 23. *An important advance in the understanding of how the brain attaches meanings to words came with the observation that words describing concrete entities activated neural systems that could be associated with the primary brain responses to those entities in a lower part of the brain (the temporal lobe). These images were made in an experiment designed to determine whether the patterns of activation in the front part of the brain (known to participate in processing language as well) also are related directly to the conceptual category of the word. Subjects were asked to recall words for animals, tools, or particular persons. The three types of words activated different areas. The activation patterns outlined in this figure show that words for tools (which may be defined by their action) activate regions also involved in movement. Such data add further support to the idea that different categories of information are represented in the brain in different ways.*

The experiment illustrated here tested whether differences in patterns of brain activation with different words reflects more specific properties of some brain regions. Antonio Damasio and his colleagues at the University of Iowa reasoned that if words were responsible for integrating activities from a variety of primary processing areas, then regions of the frontal lobe specialized for movement planning might become particularly active with words relating to action. They investigated the patterns of brain activations associated with reading words for animals, tools, and unique persons. They found that all words commonly activated the lower left side of the front of the brain, which is generally involved in language production. However, there were clear differences elsewhere. As predicted, there was activation in regions near or overlapping those in which activation is associated with movement when words for tools were presented. It is as if the brain mentally rehearses the associated action when recalling the word.

What is unknown, however, is how individual mental processes in several areas of the brain are consciously perceived as a unitary whole. Understanding the binding together of the elements of mental processing is a fundamental problem for understanding the phenomenon of human consciousness. One theory is that the complex patterns of periodic electrical activations in different areas of the brain become synchronized and that it is this synchronization of activations that links them in consciousness. Such a model suggests that multiple areas of activation work together to bind a variety of associations to a specific meaning.

PUTTING AN ENGLISH TONGUE IN A FRENCH BRAIN

It is notoriously difficult for adults to learn a second language. In Julius Caesar, *for example, a confused Casca says of Cicero's speech that "it was Greek to me." During the final act of* Henry V, *Henry attempts to woo Princess Katherine of France, despite a comically obvious language barrier dividing the two lovers. In this selection, English and French—the languages, not the armies—contend in a charming scene as the two royals battle to be understood. (Perhaps it is not surprising that Henry's eventual marriage to Katherine did not produce the desired peace between England and France!)*

Katherine: Your majesty shall mock at me; I cannot speak your England.

Henry V: O fair Katherine, if you will love me soundly with your French heart I will be glad to hear you confess it brokenly with your English tongue. Do you like me, Kate?

Katherine: *Pardonnez-moi,* I cannot tell vat is 'like me.'

Henry V: An angel is like you, Kate, and you are like an angel.

Henry V, 5.2

One of the joys of Shakespeare arises from the sense of balance in his plays. Before the tragedy of *Macbeth* becomes too unbearably intense, we have a comic interlude with the drunken porter. The sorrow in *King Lear* is tempered by the wit of

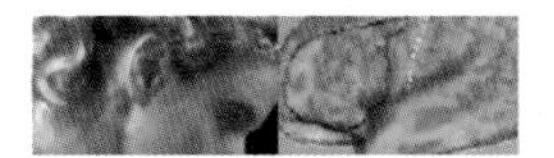

the fool. Just so, the progression of warlike patriotism in *Henry V* is broken by this charming exchange between Henry and Katherine, the young French princess. This passage provokes thoughts about important concepts of mind and language. It also raises the fascinating question of what the use of a second language means for the brain. As Henry gazes on the beautiful Katherine, he says, " . . . if you will love me soundly with your French heart I will be glad to hear you confess it brokenly with your English tongue."

O fair Katherine, if you will love me soundly with your French heart I will be glad to hear you confess it brokenly with your English tongue.

This suggests that the concept of love can be shared by both of them, even if they lack a common spoken language with which to express their feelings for each other. Shakespeare thus addresses the notion that concepts may be shared because of common modes of cognitive processing. Despite their differences, the languages of man must share underlying common fundamental mechanisms. Based on analysis of the structure of languages rather than the structure of the brain, Noam Chomsky and his school have elaborated an extreme form of this concept; they argue that there is a universal language structure that is effectively "hard-wired" in the brain. Although specific elements may differ among languages, Chomsky argues that a universal grammar exists that is shared by all languages.

In some ways, the findings of functional brain imaging and clinical observations could be said to support this hypothesis. Language processing almost always occurs predominantly in the left hemisphere, regardless of the particular language individuals use (although there may be more right hemisphere processing when reading languages—for example, Japanese kanji—that involve the use of pictographs rather than letters). Overlapping areas in the left hemisphere appear to be involved in reading, auditory comprehension, and speaking across all languages that have been studied.

Like Henry, we may refer to a foreigner's speech as "broken" English. This acknowledges an awkwardness of articulation and a distinct accent. Studies of patients who have suffered focal brain injury that has impaired their ability to generate speech have

Henry V. Vivienne Benesch as Katherine and Harry Hamlin as Henry V.

suggested that areas in the front of the brain, around what is known as Broca's area (named after the French neurologist Paul Broca, who first identified its importance) in a lower part of the left frontal brain (the inferior frontal gyrus), are critical for generating speech. Brain imaging has emphasized that activation of

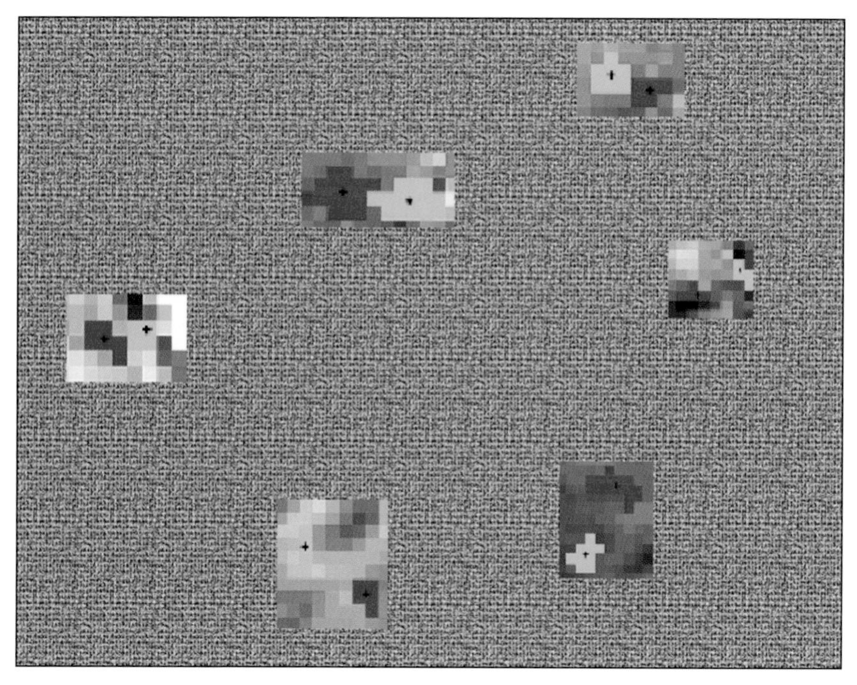

Figure 24. *These are detailed views of the activation patterns produced by an experiment in which bilingual subjects used either their native language (red) or a second language (yellow) while being monitored with fMRI. Patterns of activation were found in a small part of the lower left frontal lobe, where brain centers critical for expressive language are found. Each image is from a different individual with a unique set of first and second languages. Whether the pairing is English and French, Japanese and English, Korean and English, Spanish and English, or English and German, all of the brains showed differences in the precise localization of the areas of the cortex in this region that are responsible for the first and for the later-acquired language. Such differences in the pattern of representation of languages in the brain begin to make it possible, for example, to understand why second languages are spoken with less fluency than first languages.*

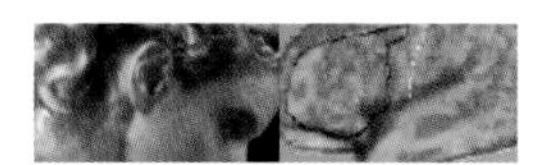

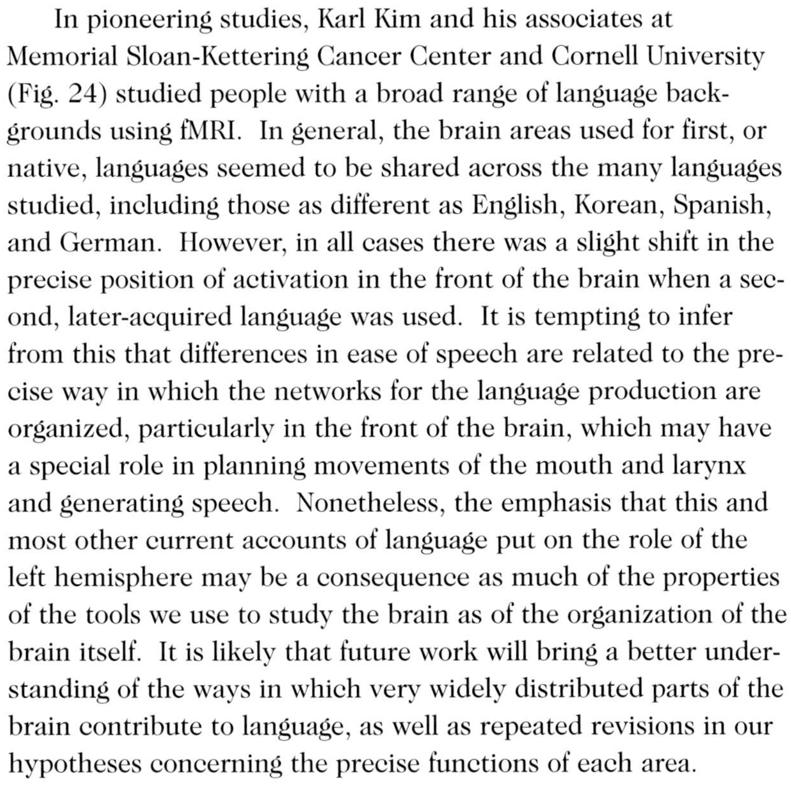

Broca's area is important for covert (or inner) speech as well as for fully articulated speech. Other experiments have been directed toward understanding what may be happening in the brains of people, such as Katherine, trying to speak a second language.

In pioneering studies, Karl Kim and his associates at Memorial Sloan-Kettering Cancer Center and Cornell University (Fig. 24) studied people with a broad range of language backgrounds using fMRI. In general, the brain areas used for first, or native, languages seemed to be shared across the many languages studied, including those as different as English, Korean, Spanish, and German. However, in all cases there was a slight shift in the precise position of activation in the front of the brain when a second, later-acquired language was used. It is tempting to infer from this that differences in ease of speech are related to the precise way in which the networks for the language production are organized, particularly in the front of the brain, which may have a special role in planning movements of the mouth and larynx and generating speech. Nonetheless, the emphasis that this and most other current accounts of language put on the role of the left hemisphere may be a consequence as much of the properties of the tools we use to study the brain as of the organization of the brain itself. It is likely that future work will bring a better understanding of the ways in which very widely distributed parts of the brain contribute to language, as well as repeated revisions in our hypotheses concerning the precise functions of each area.

The observation that *when* a person learns a language may determine exactly which part of the brain is activated while using it gives some credence to the idea of a critical period in life for acquisition of language-related motor patterns or specific grammars. This would be consistent with the concept of language being "hard-wired" in a general way, with more detailed modification of the organization of language networks by environment and use during a critical period. It would be interesting to learn whether people with a great facility for new languages show differences in the networks employed for their acquisition relative to those who find languages difficult.

NUMBERS ON THE MIND

Mathematics requires the use of mental representations of numbers to generate meaning, rather than words or letters. In Shakespeare's Julius Caesar *a meeting of conspirators immediately develops into a plot against Caesar's life. As they discuss whether Mark Antony must also die, a clock begins to strike and Brutus calls for silence to "count the clock." Some believe that this clock's chime is an anachronism, but certainly clocks existed in Caesar's day in the form of water clocks. Nonetheless, the sound of the chimes is a marvelous theatrical tool that provides a transition to drive the action forward. It also reminds us of how important a number sense is in our lives.*

Brutus: Peace! count the clock.

Cassius: The clock hath stricken three.

Trebonius: 'Tis time to part.

Cassius: But it is doubtful yet
Whether Caesar will come forth to-day or no;
For he is superstitious grown of late,
Quite from the main opinion he held once
Of fantasy, of dreams, and ceremonies.
It may be these apparent prodigies,
The unaccustom'd terror of this night,
And the persuasion of his augurers,
May hold him from the Capitol to-day.

Decius Brutus: Never fear that: if he be so resolv'd,
I can o'ersway him; for he loves to hear
That unicorns may be betray'd with trees,
And bears with glasses, elephants with holes,

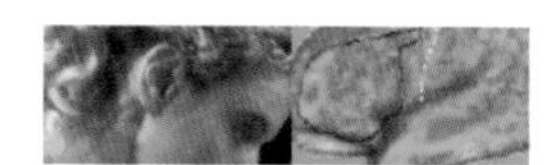

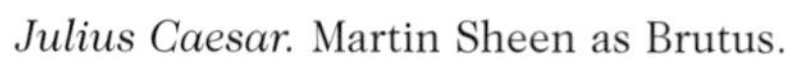

Julius Caesar. Martin Sheen as Brutus.

Lions with toils, and men with flatterers;
But when I tell him he hates flatterers,
He says he does, being then most flattered.
Let me work;
For I can give his humour the true bent,
And I will bring him to the Capitol.

Julius Caesar, 2.1

Brutus:

Peace! count

the clock.

Cassius:

The clock hath

stricken three.

By the evidence of his plays, Shakespeare's fascination for language was not mirrored by an equal interest in mathematics! However, he does make at least passing reference to some important powers of numbers. This passage opens with the chiming of a clock and its compelling demand to count out the hours. As the bells sound, Cassius is mentally counting forward to the hour as commanded by Brutus. Use of a covert or inner speech seems to be an inseparable part of such counting and even for many types of mental arithmetic.

Stanislaus Dehaene and his coworkers in France have been leading work to explore the basis of mathematical processing in the brain using the modern tools of brain science. Perhaps not unexpectedly (because working with both numbers and letters involves transforming a symbol into an idea), they have identified important parallels between number and language processing. The initial clues for these brain scientists that there are links between language processing and mathematics came from studies of patients with brain lesions affecting their mathematical abilities. In his book *The Number Sense*, Dehaene describes a Mr. N, who suffered a devastating stroke affecting the middle back part of the left side of his brain. His right hand was crippled and his ability to associate words with their meanings was impaired. From other sections of our book, you may recall that centers for language are found predominantly on the left side of the brain. Since each hemisphere primarily controls the opposite side of the body, and it is the left side of the brain that primarily controls movement on the right side of the body, this

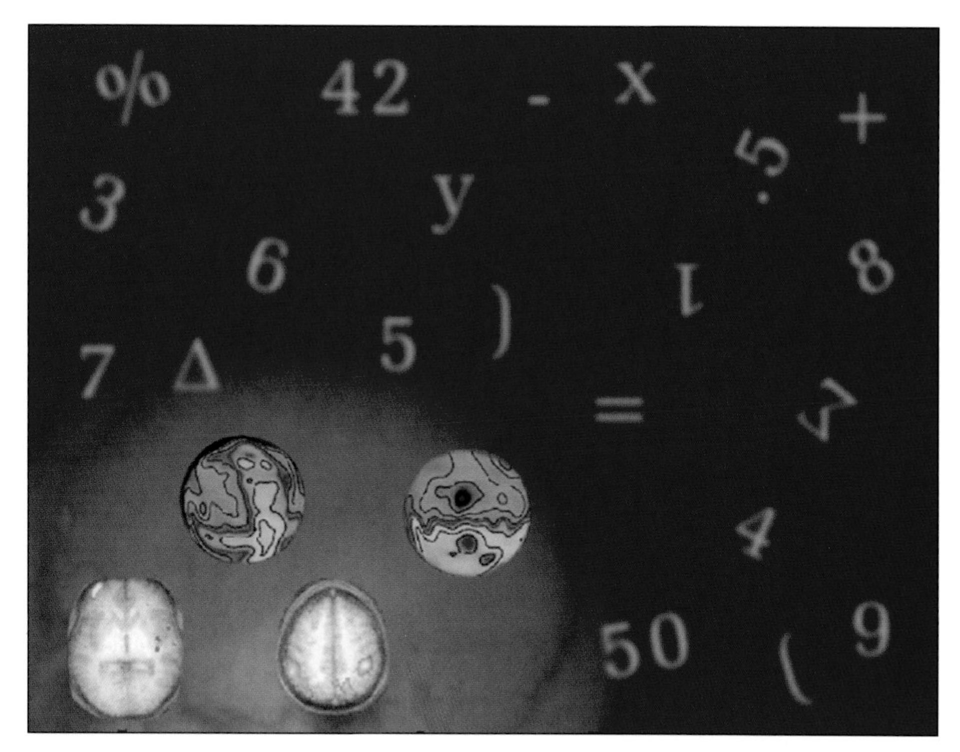

Figure 25. *The upper two images shown here illustrate differences in the pattern of electrical activity produced at the surface of the brain when a subject solved an approximate (for example, is the line on the left longer than the line on the right?) and an exact (for example, what does two plus two equal?) mathematical calculation. This "evoked potential" EEG study (during which the stimulus is repeated over and over again to allow the weak signal related specifically to the brain's response to become clear) shows that the exact calculation task (left) was associated with greater activation on the left side of the brain, while a more bilateral pattern of activation was associated with the approximate calculation (right). The lower two fMRI activation images demonstrate similar differences between the patterns of brain activity caused by these forms of calculation. Brain activation during exact calculations is found particularly on the left side of the brain, while with approximate calculations brain activity (yellow) is more equally distributed on both sides of the brain.*

accounts for the common association between weakness on the right side and difficulties with the use of language after strokes or other focal brain injuries. However, Mr. N also had a profound acalculia, or deficiency in number sense. This was so marked that when he was asked to add two plus two, he replied, after some thought, "Three."

In more recent work Dehaene has studied this phenomenon in normal individuals using both brain imaging and electrophysiological techniques (Fig. 25). When subjects were asked to perform exact numerical calculations, such as two plus two, activations were strongly lateralized to the left side of their brains, involving particularly the left inferior frontal gyrus, which is also associated with language (see "Putting an English Tongue in a French Brain," page 121) and the middle back of the brain (the supramarginal gyrus). Finding that brain representations for numbers and words are in overlapping regions of the brain may not be so surprising if one considers the many ways in which we endow numbers with meanings just as we do words—for example, 411 for "information" and 911 for "emergency."

However, people also have to deal with more general sorts of numerical problems. Caesar's superstitions are fueled in this play by a cry from the crowd to "Beware the ides of March." When we hear the date of an upcoming event, a first response is generally to classify it in some way, as, for example, either immediate, soon, or distant. In doing this we use a mental strategy similar to that for other numerical approximations, such as asking whether a book is too large to fit on a shelf or whether Chicago is halfway between New York and San Francisco. In making these rough judgments of relative proximity or magnitude, different areas of the brain are called into play from those that are used for exact calculation.

Dehaene confirmed this last notion by comparing patterns of brain activity during approximate versus exact calculation in studies of the same group of subjects. In the images shown on the lower left of the picture here, we see these differences in fMRI brain activations produced when subjects performed such

exact (blue) or approximate (yellow) mathematical calculations. In contrast to the strongly left-sided lateralization of the exact calculations, the approximate tasks showed bilateral activations, particularly in what is known as the parietal cortex, which is found in the back of the brain and is quite distinct from areas that become most active in language processing.

The parietal cortex is a so-called association or integration area that also has important functions during the performance of spatial tasks. Similar differences in brain activation between exact and approximate calculations were found by mapping the electrical patterns of brain waves, as illustrated beautifully in the images above those of the MRI brains. It is interesting that the areas involved in approximate calculations activate the back of the brain (in the parietal cortex) and thus likely overlap with those used for mental imagery, such as the images so vividly evoked by the words of Decius Brutus.

Even though we may claim to be free of superstitions, how many of us do not have a moment's pause on Friday the thirteenth? Some numbers are associated with special qualities particularly because of the associations they call forth. This characteristic of numbers may lie behind such subjective notions as the "attractiveness" of a perfect square or the "roughness" of a prime.

5. 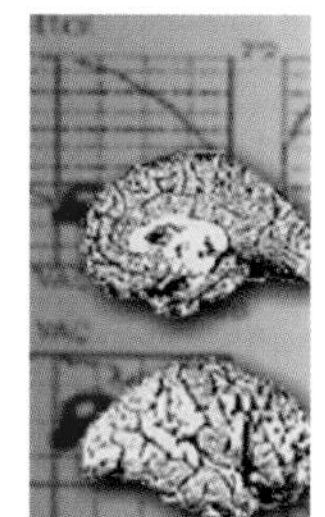 Our Inner World

DEVELOPMENT AND MEMORY

What are your earliest memories? Shakespeare explores the idea of childhood memories in The Tempest. *Prospero, once the Duke of Milan, was cruelly set adrift with his infant daughter, Miranda, in a small boat that came to land on an uninhabited island years before the action of the play begins. Now Prospero has produced a tempest to bring the old enemies responsible for his exile finally within his reach. Preparing for the confrontation, he recalls for the now adolescent Miranda their life before coming to the island. However, as he starts to do this, she surprises him by remembering images from the same time, during her earliest childhood. She recalls the women who tended her before she was three years old, and Prospero marvels at her capacity to remember anything from such an early age. He questions his daughter about her memories, and in this scene Miranda describes them as "far off,/And rather like a dream."*

The Tempest. Ana Reeder as Miranda and Ted van Griethuysen as Prospero.

Miranda:	You have often Begun to tell me what I am, but stopp'd, And left me to a bootless inquisition, Concluding 'Stay: not yet.'
Prospero:	The hour's now come; The very minute bids thee ope thine ear; Obey, and be attentive. Canst thou remember A time before we came unto this cell? I do not think thou canst, for then thou wast not Out three years old.
Miranda:	Certainly, sir, I can.
Prospero:	By what? by any other house or person? Of any thing the image tell me, that Hath kept with thy remembrance.
Miranda:	'Tis far off, And rather like a dream than an assurance That my remembrance warrants. Had I not Four or five women once that tended me?
Prospero:	Thou hadst, and more, Miranda. But how is it That this lives in thy mind? What seest thou else In the dark backward and abysm of time? If thou remembrest aught ere thou cam'st here, How thou cam'st here thou mayst.
Miranda:	But that I do not.
Prospero:	Twelve year since, Miranda, twelve year since, Thy father was the Duke of Milan, and A prince of power.

The Tempest, 1.2

Memory is one of the most remarkable and puzzling aspects of cognition. It is also among the most important to us, for a sense of personal history is critical to defining who and what we are. Miranda's choice of words is particularly appropriate

But how is it
That this lives
in thy mind?
What seest
thou else
In the dark
backward and
abysm of time?

as she tells Prospero, "You have often/Begun to tell me what I am," in reference to his recollections of her past.

Miranda has only a very imperfect memory of her early childhood. Why is it that infants and young children form only limited memories? At birth, an infant apparently lives from moment to moment. If a ball is hidden from a newborn infant's eyes by being put under a blanket, even with this done in full view, the infant behaves as if it no longer exists. Only after some months is a baby able to appreciate that the ball is still there under the blanket. Only after about 18 months of age can a toddler access memories of the past and generalize from them in order to imagine future events and their possible outcomes. It takes even longer, typically until the age of 24 months, before a child is able to realize that the world exists independently of his or her perception of it. Imagine for a moment the extraordinary concept that the world might disappear and be reformed with every blink of the eye. This highlights the problem of defining with absolute certainty what the real relationship is between our mental representations, or memory, and the "reality" of the world around us, a problem that continues to fuel philosophical debate.

To develop the mnemonic capacities that we take for granted in ourselves, some specialized mechanisms must be engaged in the brain. These develop over the years after birth. Short-term memory develops more quickly than longer-term memory. Children of two to three develop quite good "recognition" memory, which allows them to appreciate that they have seen an object before, but they still have a limited ability to recollect specific events. Recalling particular facts from long-term memory, as Miranda is attempting, is complex because it involves multiple stages of processing. The memory must be laid down or "encoded." Later, it is consolidated. Finally, when one wants to remember, the fact must be retrieved selectively from all other memories.

One of the psychologically more demanding aspects of long-term memory is this need for *selectivity* in storage as well

as retrieval. An ability to remember thus depends on processes such as being able to discriminate the important from the unimportant and learning to recognize distinguishing parameters in the frame of reference or subject area. This discrimination ability improves with life experience and therefore is relatively slow to develop. To some extent, it must be learned afresh for each subject area. Thus, when Prospero asks whether Miranda remembers the time before they came to their lonely island, he quickly adds, "I do not think thou canst, for then thou wast not/Out three years old."

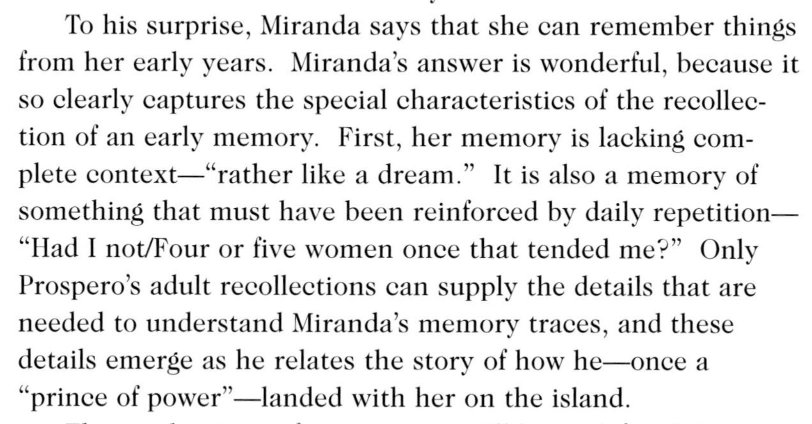

To his surprise, Miranda says that she can remember things from her early years. Miranda's answer is wonderful, because it so clearly captures the special characteristics of the recollection of an early memory. First, her memory is lacking complete context—"rather like a dream." It is also a memory of something that must have been reinforced by daily repetition—"Had I not/Four or five women once that tended me?" Only Prospero's adult recollections can supply the details that are needed to understand Miranda's memory traces, and these details emerge as he relates the story of how he—once a "prince of power"—landed with her on the island.

The mechanisms of memory are still being defined, but it is clear that they involve complex networks that include widely separated parts of the brain. Specific areas seem to be particularly important to all memories for events and places, such as the structures in the middle part of the brain known as the hippocampus and the parahippocampus. Neurons in these regions appear to integrate activity from widely distributed regions of the brain. The need for functional integration of activity across the brain makes the connections between neurons (the myelin-coated axons) as important as the neurons themselves.

Differences in the quality of memory formation between young children and adults probably result from many factors. However, developmental changes in the brain that affect the ways in which individual cells interact and the overall structure of the brain must be very important indeed. Unlike many other organs of the body that simply grow in size as a child develops, the brain changes qualitatively, with different rates of growth in different parts.

The brain grows most dramatically before birth. This growth occurs particularly in the cortex, the surface of the brain that includes nerve cell bodies. To accommodate the

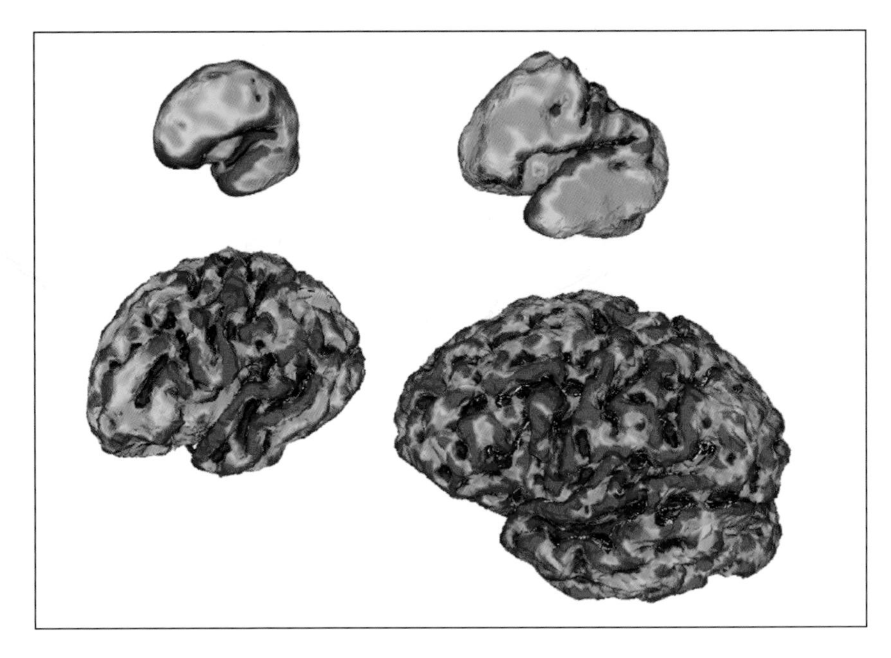

Figure 26. *One of the extraordinary features of the human brain is the large size of the cortex, the surface layer of nerve cells responsible for most "higher" brain activities. The cortex becomes folded during brain development to fit its large surface area inside the skull. This figure shows images of the fetal brain at 23, 26, 36, and 40 weeks of gestation, over which time complete folding occurs. The colors (with red showing the curvature of a surface "peak" or gyrus and blue showing the curvature of a surface "valley" or sulcus) emphasize the development of the folds in the cortex as the brain matures. The initial increase in size of the cortex comes as new nerve cells migrate into it. However, the increases in area (and thus the amount of folding) that occur late in gestation are due more to increased branching and formation of integrating connections between nerve cells than to development of new nerve cells in the cortex.*

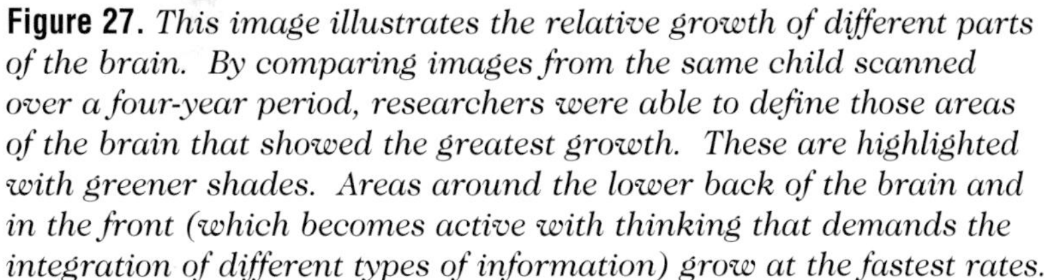

Figure 27. *This image illustrates the relative growth of different parts of the brain. By comparing images from the same child scanned over a four-year period, researchers were able to define those areas of the brain that showed the greatest growth. These are highlighted with greener shades. Areas around the lower back of the brain and in the front (which becomes active with thinking that demands the integration of different types of information) grow at the fastest rates.*

large number of cells and their connections without expanding the total brain size to the point that a baby's head could not squeeze through the birth canal, the cortex (which if flattened would be as large as a dish towel) develops the folded gyri that are so characteristic (Fig. 26). Growth continues through childhood and into adolescence.

These structural changes can be mapped in some detail using modern brain imaging. For example, the growth in early childhood of areas of the brain that are involved in the integration or association of ideas is illustrated by this picture from Paul Thompson's laboratory at the University of

California at Los Angeles (Fig. 27). It shows patterns of growth (green color) in the brain after birth in greater detail than has previously been possible. Two representations of relative brain size changes are shown. They were created by comparing a brain imaged at two different times and then representing local increases in size by lighter color. The lower representation shows that changes in brain size over a period of two weeks were negligible, as indicated by the uniformity of the color. The upper representation shows large patches of the lighter color. Over a four-year interval there was marked growth, particularly in the back and lower parts of the brain, where so-called association cortex (responsible in part for integration and manipulation of information from primary sensory areas of the brain) is found. The overall size of the brain changes during this time as well. However, to make the regional differences clearly apparent, Thompson used computational techniques to make all the brains assume a common size. Measuring the relative rates of change of different areas of the brain helps us understand the basis for changes in mental capacities with time. These data tell us that as the child matures, areas particularly of "association" cortex not highly specialized for primary cognitive processing tasks enlarge in parallel with the developing capacity for abstract thought.

MUSIC AS A CALL TO LIFE

Music seems to have a magic all its own. Shakespeare recognizes this magic and dramatizes its potency in one of his later plays, The Winter's Tale. *Dramatic intensity is focused on the final scene of the play, in which music appears to bring a statue to life. At the beginning of the play, after the Sicilian king Leontes' jealous rage against her, his innocent wife, Hermione, wrongfully imprisoned for adultery, conspires to be reported dead. With this following the death of his only son, the lonely king repents his anger. Paulina, a faithful attendant of Hermione, then unveils a "statue" that proves to be a remarkably exact replica of the supposedly dead queen. In the scene that follows, after Paulina reawakens Leontes' faith in his wife, her demand for music acts as a seemingly magical call to life for the "statue."*

Paulina: Music, awake her; strike!

[Music.]

'Tis time; descend; be stone no more; approach;
Strike all that look upon with marvel. Come!
I'll fill your grave up: stir, nay, come away:
Bequeath to death your numbness; for from him
Dear life redeems you. You perceive she stirs:

[Hermione comes down.]

Start not; her actions shall be holy as
You hear my spell is lawful.

[to Leontes] Do not shun her
Until you see her die again; for then
You kill her double. Nay, present your hand:
When she was young you woo'd her; now, in age,
Is she become the suitor?

The Winter's Tale. (Left to right) Mandy Patinkin, Alfre Woodard, James Olson, Diane Venora, Christopher Reeve, Jennifer Dundas and Graham Winton.

Leontes: O, she's warm!
If this be magic, let it be an art
Lawful as eating.

The Winter's Tale, 5.3

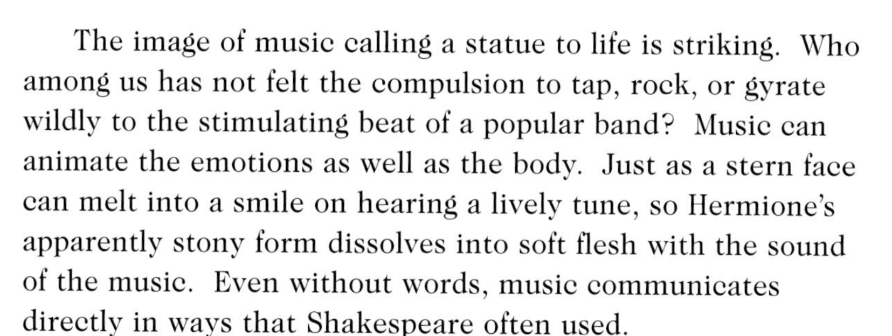

The image of music calling a statue to life is striking. Who among us has not felt the compulsion to tap, rock, or gyrate wildly to the stimulating beat of a popular band? Music can animate the emotions as well as the body. Just as a stern face can melt into a smile on hearing a lively tune, so Hermione's apparently stony form dissolves into soft flesh with the sound of the music. Even without words, music communicates directly in ways that Shakespeare often used.

Music,
awake her;
strike!

Shakespeare's plays are full of music, and he very consciously employs its power to move his audience. Songs such as the one at the end of *Twelfth Night*—"When that I was and a little tiny boy/With hey, ho, the wind and the rain,/A foolish thing was but a toy,/For the rain it raineth every day"—provide an appropriate denouement, or rounding out of the action, at the close of an act. There also is often a real sense of music in the rhythm of Shakespeare's words and the timbre of their sounds.

Modern neuroscientists now appreciate that musical processing activates many parts of the brain. Research is being conducted in several laboratories to define precisely how music is perceived in the brain. Richard Frackowiak and his colleagues at the Functional Imaging Laboratory of University College in London are among those who performed some of the first neuroimaging studies of the effects of music on the brain. In their experiments using PET, they varied specific qualities of music for a group of listeners. By recording brain activity while changing pitch and rhythm independently, for example, the images shown here were generated (Fig. 28; overleaf). They demonstrate that processing of these specific musical elements is performed by large groups of nerve cells primarily

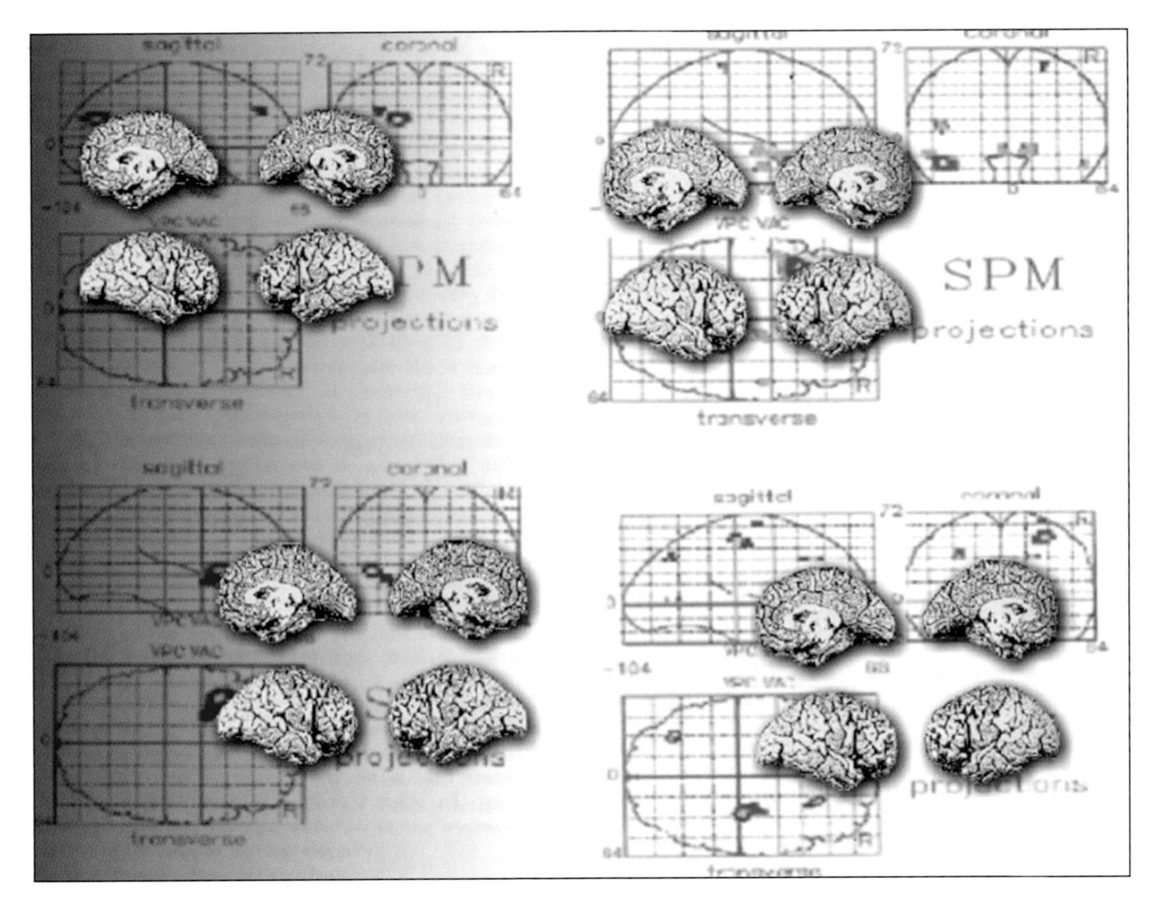

Figure 28. *These PET images illustrate ways in which the basic elements of music are processed in the brain. Subjects were asked to listen to a tape of a sequence of randomly arranged notes. In different parts of the study, they were asked to respond specifically to aspects of familiarity, rhythm, pitch, and timbre. The images show differences in brain activity for a range of contrasts: pitch relative to timbre and rhythm (upper left), timbre relative to pitch and rhythm (lower left), familiarity relative to pitch (upper right), and pitch relative to rhythm (lower right). In general, the left side of the brain was more active for tasks that involved judgments of familiarity, pitch and rhythm, while the right hemisphere was more active for judgments of timbre.*

found in the left side of the brain. There appears to be considerable overlap with areas of the brain that are used for language. For example, both language and rhythm activate the so-called superior temporal gyrus, the fold on the top of the lower lobe of the brain, and Broca's area, in the lower part of the front of the brain. This should probably not be surprising, as making sense of the sounds of words also must involve an appreciation of rhythm and pitch. One of the greatest difficulties with learning a foreign language, for instance, is appreciating the typical patterns of sounds that define words and sentences in that language.

In their studies of music, Frackowiak and his colleagues were surprised to find that variation of pitch discrimination primarily activated areas in the back of the brain that had previously been thought to be used primarily for visual perception. One possible explanation for this is that pitch decoding may involve a visual component. Mental imagery may be used in some way to encode patterns of relative pitch in music. If this is true, then when we speak colloquially of a musical pitch as being "higher" or "lower," it may be a reflection of a bias in the human brain for processing relationships in spatial terms.

Frackowiak's experiments also identified brain activity in the back part of the right side during judgments of timbre. This area (known as the right parietal lobe) is activated also (although obviously not exclusively) for tasks that demand thinking about spatial relations. Timbre may in some respects require a type of thinking analogous to spatial processing, because it involves understanding the relationships between different pure tones that make up the sound as a whole.

Together these observations illustrate how much of the brain is involved in the processing of music—suggesting how completely it can dominate our consciousness. Music truly calls the brain as a whole to life!

WHAT MAKES MUSIC SWEET?

The brain's appreciation of the sweetness of music plays an important role in many of Shakespeare's comedies, and Twelfth Night *is no exception. In fact,* music *appears in the opening line, spoken by the sentimental Orsino, duke of Illyria, as he indulges himself in lovesickness. He vainly seeks the love of the countess Olivia. Eventually Orsino comes to realize that the shipwrecked Viola is his true love, while Olivia falls in love with Viola's twin brother, Sebastian. Until that time, however, the duke is forced to try to calm his romantic longings by clever wordplay and filling his ears with music— "the food of love."*

Orsino: If music be the food of love, play on,
Give me excess of it, that, surfeiting,
The appetite may sicken, and so die.
That strain again, it had a dying fall:
O, it came o'er my ear like the sweet sound
That breathes upon a bank of violets,
Stealing and giving odour. Enough, no more;
'Tis not so sweet now as it was before.
O spirit of love, how quick and fresh art thou,
That notwithstanding thy capacity
Receiveth as the sea, nought enters there,
Of what validity and pitch soe'er,
But falls into abatement and low price,
Even in a minute! So full of shapes is fancy,
That it alone is high fantastical.

Twelfth Night, 1.1

For Orsino, love represents life's greatest pleasure, and its absence is one of the sources of greatest pain. He tries to distract himself from the melancholy of unrequited love with music.

Twelfth Night. Michael Rudko as Feste and Graham Winton as Orsino.

If music be the
food of love,
play on,
Give me excess
of it, that,
surfeiting,
The appetite
may sicken,
and so die.

Although the sounds may have changed, music is used in the same way by unhappy lovers in our own times. Here, Shakespeare builds on the universality of this experience.

There is interest for the brain scientist in considering the ways in which Orsino links music and love: "If music be the food of love, play on." Shakespeare is observing both that music can contribute to the atmosphere of love and that the outwardly very different experiences of love and music evoke a common feeling of pleasure.

Deciding precisely what we mean by pleasure is difficult, although it is easy to recall experiences that gave us pleasure. One attempt at a concise definition is to say that pleasure is a reward state that reinforces behaviors. It is an experience that can be evoked equally by different stimuli and situations. Orsino calls for his musicians to replay a song which, he says, " . . . came o'er my ear like the sweet sound/That breathes upon a bank of violets." He hopes that he can generate pleasure from the music that in some way might compensate for the loss of pleasure from love.

Orsino's pleasure in love is focused specifically on Olivia, and he is blind to the potential pleasures of (the disguised) Viola, who is about to make her appearance. Similarly, it is a common experience that the same music does not give pleasure to everyone. Pleasure is not invariably associated with a given type of experience (or stimulus) but depends on its quality and context and our learned expectations. For example, few Western listeners find classical Chinese music soothing (or even comprehensible!) on first hearing it, despite the pleasure many Chinese listeners take in it.

Love is a particularly difficult emotion for modern neuroscience to explore because it is so complex. However, observations of fundamental psychological characteristics such as those made by Shakespeare in this quotation from *Twelfth Night* can help guide brain scientists in developing approaches to understanding the basis for feelings of pleasantness and unpleasantness that are common to love, music, and many other experiences.

Differences in the ways that the brain works when listening to pleasant or unpleasant music have been studied using PET by Robert Zatorre and his colleagues at the Montreal Neurological Institute (Fig. 29). To do this, they played a series of musical

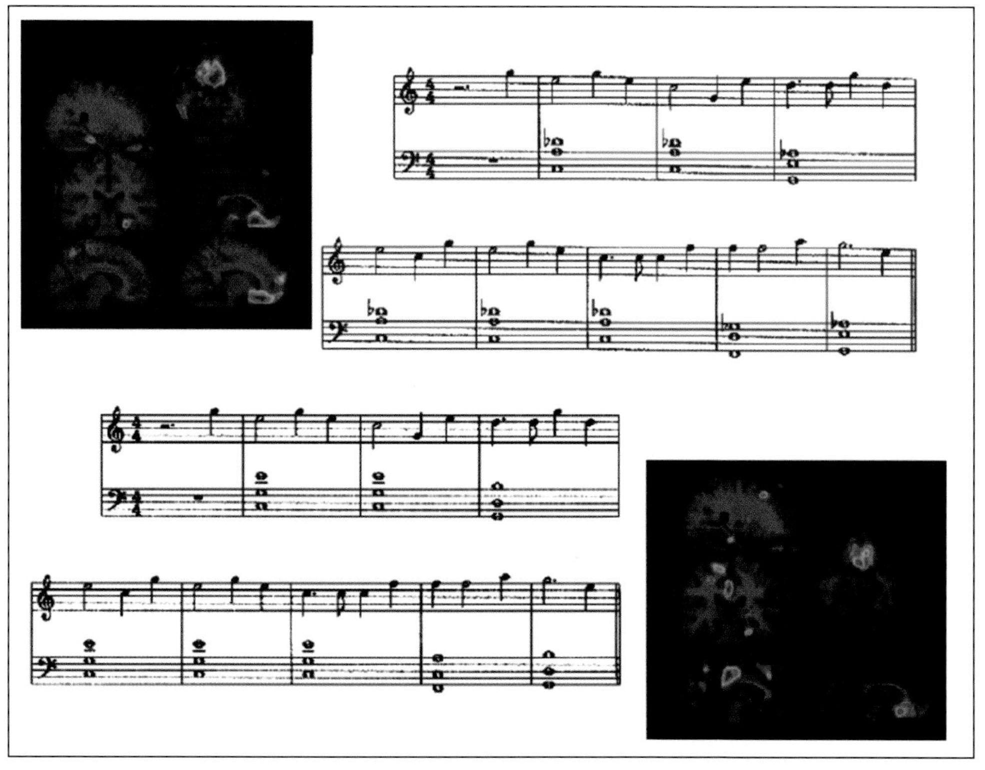

Figure 29. *To generate these PET images, subjects were asked to listen to brief selections of music. The images shown in the upper left were made by correlating brain activity directly with the relative degree of musical dissonance. Bright orange defines those areas that showed the greatest increase in activity with greater dissonance. In contrast, the blue areas are those that showed decreases in activity with greater dissonance. Striking activation at the base of the front of the brain is seen in an area that has been shown to be involved in responses to both pleasant and unpleasant stimuli. The images in the lower right illustrate this idea more directly by demonstrating the correlation of patterns of brain activation with the pleasantness or unpleasantness of the sounds as rated by the listeners. They also show activation at the base of the front of the brain.*

passages to their subjects that varied in dissonance and therefore in the perceived degree of pleasantness. They then distinguished those areas of the brain that became more active when the more pleasant passages were played. They found that these included parts of the base of the front of the brain, in what is known as the orbitofrontal cortex (because it lies above the eye sockets, or orbits). They also found activation in a central region of the brain known as the cingulate cortex.

What is exciting about finding activation in the orbitofrontal cortex is that the same part of the brain has been shown by Edmund Rolls and his colleagues at Oxford to be activated by a pleasant feeling of touch, such as from stroking soft velvet. It thus appears as though the common experience of pleasure evokes activity in the same or similar areas of the brain regardless of the specific nature of the trigger. This is precisely what Orsino hopes for when he refers to music as "the food of love."

Of course, Orsino is also attempting to use the music to cure him of his lovesickness. He cries, "Give me excess of it, that, surfeiting,/The appetite may sicken, and so die." What he is relying on is that too much of even a pleasurable stimulus brings satiety (and, subsequently, even disgust). However—and unfortunately for Orsino—more recent research by Rolls and his team has confirmed that a sense of satiety for one previously pleasant stimulus does not necessarily bring satiety for another. The plot of *Twelfth Night* must thus proceed, and love must be requited before the pain of absent love can be assuaged.

BEING GOD'S SPY: OPENING THE EMOTIONAL HEART

The ability to express and discuss our emotions is an important part of what makes us human. At the end of King Lear, *the old king and his faithful daughter Cordelia are reunited but then captured by enemy forces. Even these difficult circumstances do not mar the happiness of their reconciliation. Lear suggests that they forgo meeting his other two daughters, who have betrayed him, and be content to go to prison to be "like birds i' the cage." There, he says, they will retain the freedom to think despite physical confinement. They will also have an opportunity to savor the joy that trading blessings and forgiveness with each other will bring. Lear has finally become able to see the true love of his youngest daughter. In the exhilaration of his enlightenment, he describes Cordelia and himself as "God's spies"—observers of the human condition who are able to see directly what is written in the hearts of their fellow men.*

Cordelia: We are not the first
Who with best meaning have incurred the worst.
For thee, oppressed King, I am cast down;
Myself could else outfrown false fortune's frown.
Shall we not see these daughters and these sisters?

King Lear: No, no, no, no. Come, let's away to prison;
We two alone will sing like birds i' the cage.
When thou dost ask me blessing I'll kneel down
And ask of thee forgiveness. So we'll live
And pray, and sing, and tell old tales, and laugh
At gilded butterflies, and hear poor rogues
Talk of court news; and we'll talk with them too—
Who loses and who wins, who's in, who's out—

King Lear. Monique Holt as Cordelia and Ted van Griethuysen as Lear.

And take upon's the mystery of things
As if we were God's spies. And we'll wear out
In a walled prison packs and sects of great ones
That ebb and flow by the moon.

King Lear, 5.3

So we'll live
And pray, and
sing, and tell
old tales, and
laugh
At gilded
butterflies....

The reconciliation of Cordelia with Lear is one of the most moving scenes in this powerful play. The deep emotions of the scene raise a number of intriguing questions for brain science. Cordelia opens the passage by referring to the way in which she might "outfrown false fortune's frown," reminding us of the key role facial expressions play in communicating our emotions.

Until relatively recently, we understood little about the biological basis for the perception of emotions. Much of the playwright's art lies in being able to manipulate the emotions of the audience, and an audience enjoys the communication of emotions projected by a skilled actor in character. The popular film *Shakespeare in Love* suggests that the passion in Shakespeare's plays could have been a direct result of passion in the playwright himself at the time of writing—lust fueling words of love (although it is hard to believe that even the most committed playwright would interrupt everything to put pen to paper!). That miming an emotion, or watching another do so, can actually cause one to feel the emotion strongly reminds us that there is a close relationship between the body's responses to emotion and its mental experience. In fact, the famous nineteenth-century American psychologist William James suggested that the physical responses of the body to an emotional state (such as the feeling of cold sweat, the increase in heart rate, and the tightening of the stomach associated with fear) were what generated the mental experience of the emotion.

We do not believe this now. Rather, current thinking holds that emotions are generated solely in the brain. A guiding concept is that activity in a central "circuit" is associated with emotion. This hypothesis was first put forward by the Portuguese brain scientist Igor Papez, who postulated that specific areas of the middle part of the brain (in what is now known as the limbic system) are responsible for both the conscious and unconscious experience of emotional states. One of the most complex parts of this circuit is a small almond-shaped structure near the middle front of the brain, called the amygdala. The amygdala turns out to be critically important to our appreciation of the emotional experiences of others.

Neuroimaging studies performed by Ray Dolan and his colleagues in the Functional Imaging Laboratory of University College in London have demonstrated, for example, that the perception of emotional meaning in a frowning or angry face activates the amygdala. In one study, people were briefly shown faces that had either fearful or neutral expressions (Fig. 30).

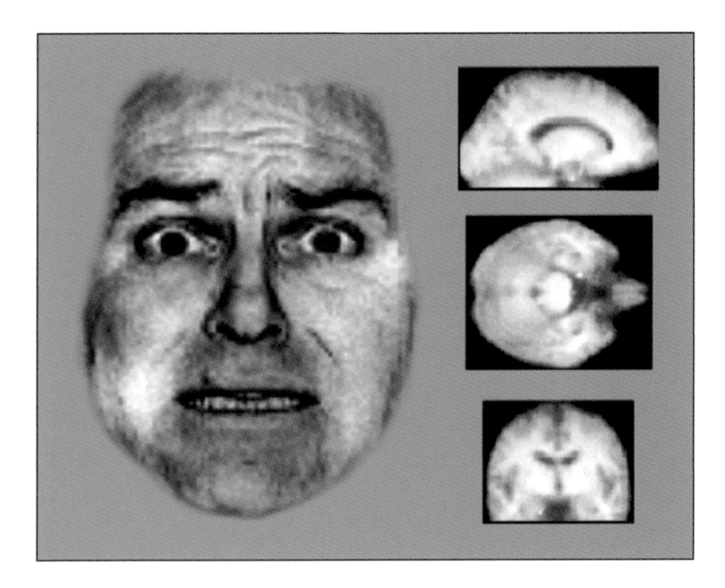

Figure 30. *To generate the PET images shown here, subjects were asked to view a series of faces showing increasingly fearful expressions. Viewing a fearful facial expression specifically activated the left amygdala (yellow), a small, almond-shaped brain structure that is part of the limbic system, or "emotional brain." The amygdala may have an important role in forming memories of emotional situations.*

Then they were immediately shown another such image, but for a longer period of time. The second image distracted the subjects from a conscious experience of the first image. Nevertheless, despite the lack of conscious experience of the first images, careful measurements of the amount of perspiration in the skin and the dilation of the pupils (which both increase when a threatening facial expression is perceived) showed that subjects subconsciously responded to the first image, providing an indirect index of emotional response. PET imaging showed an increase in activation of the amygdala after the frowning faces were shown, but not after faces with neutral expressions. The emotional content of a facial expression therefore is processed in the amygdala independently of its conscious experience.

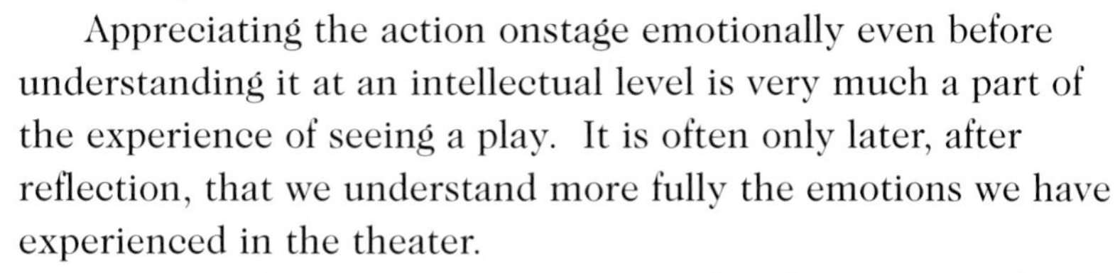

Appreciating the action onstage emotionally even before understanding it at an intellectual level is very much a part of the experience of seeing a play. It is often only later, after reflection, that we understand more fully the emotions we have experienced in the theater.

Lear is perhaps commenting on a similar phenomenon when he describes to Cordelia the pleasures they might have if they were to contemplate the world outside of the imagined leisure of prison. As they talk to "poor rogues" from the court, they will be able to "take upon's the mystery of things/As if we were God's spies." This ability to perceive the emotional state of others directly (sometimes without a need for spoken language) is a quality that may seem at times to confer an almost godlike omniscience.

A FEARFUL ANTICIPATION

Arrogant in his belief in the witches' prophecy that he cannot be killed by anyone "of woman born," Macbeth has lived without fear except from the demons of his own mind. However, in the scene from which the quotation below is taken, Macduff reveals that he was delivered by a form of cesarean section (named, incidentally, for Julius Caesar, who was said to be born the same way). Macbeth is briefly shocked by the realization that the erstwhile comrade-in-arms facing him is therefore capable of killing him. In this moment of anxious fear, when he is "cow'd," Macbeth realizes the cruel joke of the half-intelligence offered by the witches, who he refers to as "juggling fiends."

Macbeth: Thou losest labour:
As easy may'st thou the intrenchant air
With thy keen sword impress, as make me bleed:
Let fall thy blade on vulnerable crests;
I bear a charmed life; which must not yield
To one of woman born.

Macduff: Despair thy charm;
And let the Angel, whom thou still hast serv'd,
Tell thee, Macduff was from his mother's womb
Untimely ripp'd.

Macbeth: Accursed be that tongue that tells me so,
For it hath cow'd my better part of man:
And be these juggling fiends no more believ'd,
That palter with us in a double sense;
That keep the word of promise to our ear,
And break it to our hope.—I'll not fight with thee.

Macbeth, 5.8

Macbeth. Stacy Keach as Macbeth and Chris McKinney as Macduff.

Accursed be that tongue that tells me so, For it hath cow'd my better part of man....

An unsettled world is not a modern phenomenon. Shakespeare knew anxiety as keenly as we do and expected his audience to identify with Macbeth's sudden fear easily. In this confrontation with impending pain and death, Macbeth would feel immobilizing weakness, "butterflies" in his stomach, sweating of the palms, and palpitations of the heart. These all result from activation of the autonomic nervous system, which is part of the body's unconscious control system for the basic physiological functions (such as temperature, heart rate, and blood pressure) necessary for survival. Activation of the autonomic nervous system prepares the body for immediate action. However, when this occurs in a situation in which successful action is impossible, these changes contribute to feelings of anxiety. Macbeth realizes the impossibility of triumphing over Macduff and vainly tries to ward off an inevitable fate, crying, "I'll not fight with thee!"

As noted in the last section, nineteenth-century psychologists, led by William James, believed that the autonomic responses themselves fed back into the brain to generate the conscious experience of emotions such as anxiety. A more modern view, however, is that the activity of the brain is directly responsible for the conscious experience of anxiety and that this activity feeds forward to control the body's physiological changes by regulation of the autonomic nervous system. This notion is now tested routinely by people who use beta-blocker drugs, like the popular blood pressure drug propranolol, which can blunt the autonomic responses (such as sweating palms) but clearly do not reduce the subjective mental experience of anxiety.

A major challenge to modern neuroscience has been to understand the ways in which higher brain centers develop emotions such as anxiety. A recent experiment that has provided some insight into this is illustrated here. Decades ago the pioneering Russian physiologist Ivan Pavlov provided a model for a special type of learning called conditioning in studies of dogs. Hungry dogs always begin to salivate at the sight of food, which therefore is called an unconditional stimulus. Pavlov found that his dogs could be trained to salivate at the sound of a bell simply

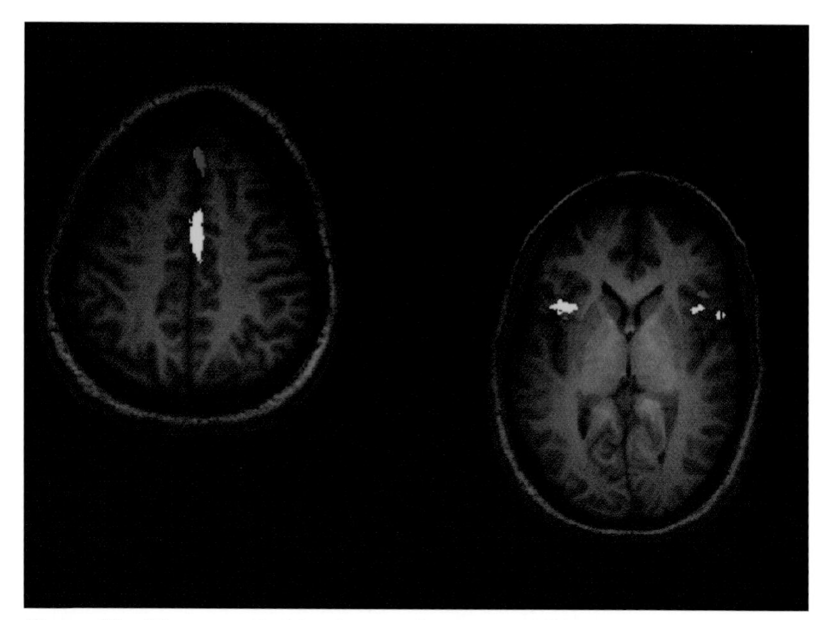

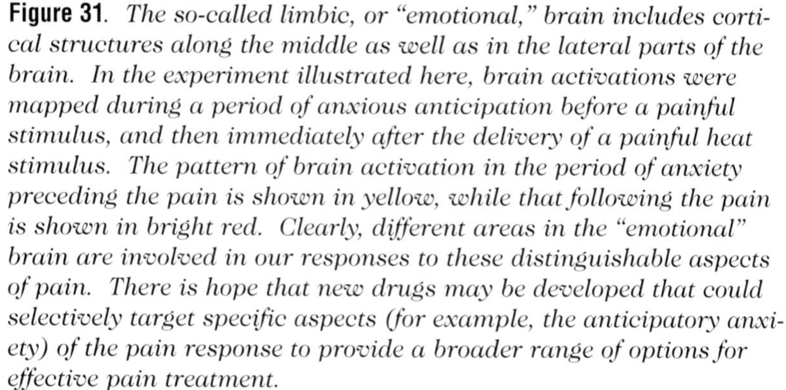

Figure 31. *The so-called limbic, or "emotional," brain includes cortical structures along the middle as well as in the lateral parts of the brain. In the experiment illustrated here, brain activations were mapped during a period of anxious anticipation before a painful stimulus, and then immediately after the delivery of a painful heat stimulus. The pattern of brain activation in the period of anxiety preceding the pain is shown in yellow, while that following the pain is shown in bright red. Clearly, different areas in the "emotional" brain are involved in our responses to these distinguishable aspects of pain. There is hope that new drugs may be developed that could selectively target specific aspects (for example, the anticipatory anxiety) of the pain response to provide a broader range of options for effective pain treatment.*

by ringing it every time they were fed. With time, the relationship between salivation and the sound of the bell became independent of actually giving the dogs food. In such a situation the salivation is known as a conditional response, and the ringing of the bell a conditional stimulus. A similar type of experiment can be performed with people.

To generate the images shown in Figure 31, Oxford scientists Alex Ploghaus and Irene Tracey studied subjects with functional magnetic resonance imaging (fMRI). The subjects were shown colored lights. A green light was followed by the application of a rather pleasant, mild heat stimulus through a metal plate strapped to the back of the hand. But, following a delay during which a red light was shown, a blue light could appear and the metal plate would become painfully hot (although never enough to cause any lasting harm). Subjects rapidly learned the association between the colors of the lights and the imminence of pain. The patterns of activation in the brain after subjects saw each color could be studied separately. By analyzing the pattern produced by the red light, the scientists could selectively assess brain responses to the anxiety preceding a painful stimulus.

The first gratifying result was that the responses during the anxious period before the pain were different from those during the painful stimulus itself. During this anxious phase, areas that were activated particularly involved those parts of the brain known as the limbic system, the part of the brain whose activity is closely linked to motivation and emotion. Although he would not have been aware of it, Shakespeare's choice of plots, literary techniques, and stagecraft all manipulate activity in this region of the brains of his audience.

Just as Shakespeare's poetry can excite or calm anxiety, fear, loneliness, love, and other emotions, so a major goal of current psychopharmaceutical studies is to develop ways of better controlling undesirable or pathological feelings. An exciting possibility suggested by the studies described here is that it may be possible to use brain imaging to test new drugs and develop safer and more selective therapies for anxiety. Ready availability of such agents would not be without concerns, however. Feelings such as anxiety can act to keep us aware of dangers and the potential consequences of our actions, even when we consciously try to suppress such thoughts. Did Macbeth's downfall begin when his fear and anxieties about the future were removed by the witches' words on the heath? Is Shakespeare warning us about the dangers of removing a deep-rooted appreciation for the link between actions and their consequences?

BUILDING A WORLD IN THE MIND

Through the imagination, an entire world may be constructed in the human mind, and such a vision is developed onstage in one of Shakespeare's final plays, The Tempest. *Offering some of the playwright's most enchanting poetry,* The Tempest *includes a supernatural spectacle enacted by spirits. It is a fantasy world, a premarital pageant created by the wizardry of Prospero to entertain his daughter, Miranda, and her newfound love, Ferdinand. Remembering the conspiracy of the evil Caliban, Prospero dismisses the spirits and explains to his future son-in-law the illusory nature of this created vision. By extension, Prospero's words regarding his own magic apply equally well to any visions of the "mind's eye."*

Prospero: You do look, my son, in a mov'd sort,
As if you were dismay'd: be cheerful, sir.
Our revels now are ended. These our actors,
As I foretold you, were all spirits, and
Are melted into air, into thin air:
And, like the baseless fabric of this vision,
The cloud-capp'd towers, the gorgeous palaces,
The solemn temples, the great globe itself,
Yea, all which it inherit, shall dissolve,
And, like this insubstantial pageant faded,
Leave not a rack behind. We are such stuff
As dreams are made on; and our little life
Is rounded with a sleep. Sir, I am vex'd;
Bear with my weakness; my old brain is troubled:
Be not disturb'd with my infirmity:
If you be pleas'd, retire into my cell,
And there repose: a turn or two I'll walk,
To still my beating mind.

The Tempest, 4.1

The Tempest. Claudie Blakely as Miranda, Ian McKellen as Prospero, and Rashan Stone as Ferdinand.

Shakespeare must have seen and heard his plays inside his head before they were ever on stage. Like Mozart, who claimed that when writing music he was simply transcribing what he heard in his mind, Shakespeare must have imagined his plays with what a character in *A Midsummer Night's Dream* calls a "poet's eye, in a fine frenzy rolling." Using mental imagery, he would have been able to establish "forms of things unknown" so that his "poet's pen/Turns them to shapes, and gives to airy nothing/A local habitation and a name." We use a similar creative process when our imagination is triggered to regenerate these images by the sounds of Shakespeare's words. This is acknowledged explicitly in the Prologue to *Henry V* when the Chorus asks the audience to "let us, ciphers to this great account,/On your imaginary forces work." Mental imagery, fundamental to the work of Shakespeare, clearly fascinated him.

We are such stuff As dreams are made on; and our little life Is rounded with a sleep.

Just as mental imagery forms a foundation for art, Shakespeare reminds us that it is very much a part of our inner life. Our imagination can be triggered by chance associations—a scrap of music, a place-name, or a smell, for example. Marcel Proust famously refers to this at the beginning of his multivolume novel sequence, *Remembrance of Things Past*, in which all the recollections that the story relates are said to have been unleashed by the sweet smell of a madeleine, a kind of small French cake. The associations triggered for us by our sensory experiences make critical contributions to the way in which we view the world. As is described elsewhere in this book, the patterns of brain activations associated with the recall of concrete words, for example, involve areas of the brain that contribute to the primary sensory experiences of the objects to which those words refer.

Although initiated by sensory inputs, sensory perception is not passive—it is very much a cognitive process. For particular types of sensory inputs, the mental context of the input may have a dominant role in determining the quality or significance of the type of perception. For example, consider the way in which we respond to unpleasant stimuli. When we are happy and unconcerned in a nonthreatening environment, a mildly noxious stimulus can be merely irritating. However, when associated with depression, stress, or fear, the same stimulus can be deeply unpleasant. A noxious odor that appears inescapable or uncontrollable can often be particularly unpleasant.

Shakespeare wrote, "Lovers and madmen have such seething brains," recognizing an important difference between the mental imagery an artist consciously uses and that which we involuntarily employ in perception or that which can figure in psychopathology. The artist employs imagery under the

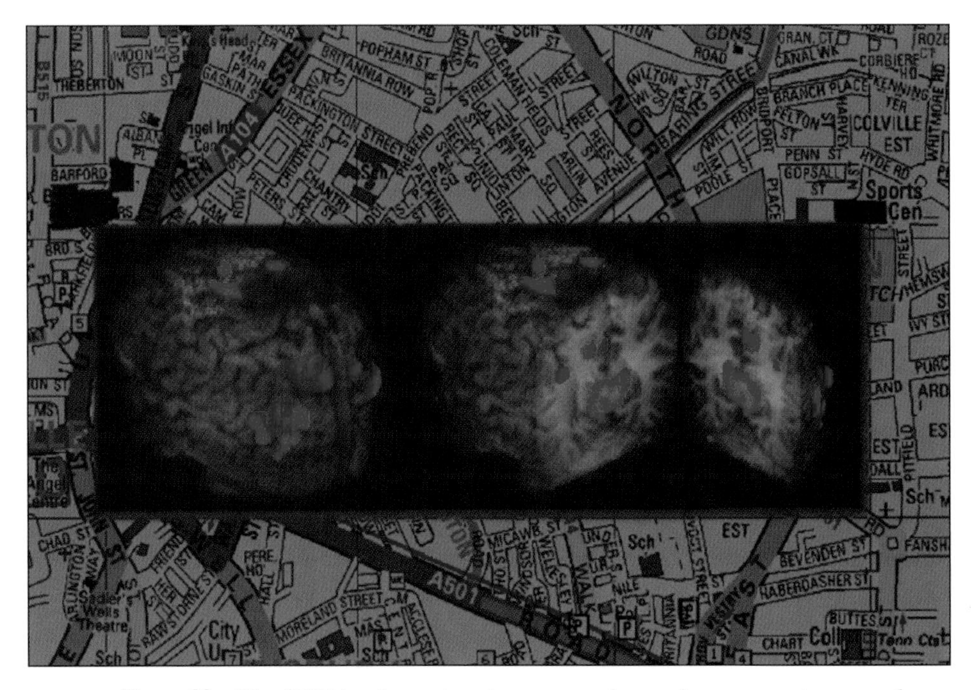

Figure 32. *The fMRI brain activation scans shown here were generated as subjects imagined that they were walking through the streets of their hometowns. This task involved extensive use of visual imagery as subjects recalled specific features of streets, as well as the directions needed to go from one point to the next. What is most striking is that areas of the brain activated during the mental imagery task were similar to those that would be involved in the processing of visual information from a real walk.*

control of reason. In contrast, the mental imagery accompanying perception is initially unconscious and therefore more difficult to separate from the experience itself.

The pathological mental imagery that accompanies some forms of brain dysfunction can be particularly difficult to distinguish from the triggering stimuli. Delusions are abnormal mental images triggered by sensory input to a depressed or otherwise altered brain. Delusions may occur after surgery, for example, when patients are not fully alert because of the stress of surgery and the drugs administered during it. These mental images are generated as the brain tries to form a meaningful whole with sensory information that is incomplete or inadequately integrated because of brain dysfunction from drugs or injury. Although the causes can be diverse, such delusional images can be highly stereotyped. Affected patients frequently describe crawling insects or water dripping down the hospital walls. Characteristic of these experiences is that they are vivid and that they are not recognized clearly by the patients themselves as being inaccurate representations of the external world.

Brain imaging studies are now starting to try to define the mechanisms of mental imagery in the healthy brain. Research has begun with attention on the basic processes that are activated when we imagine things, particularly those involved with visual imagery. Kamil Ugurbil and his colleagues at the University of Minnesota were among the first to explore this. They used fMRI to study the patterns of brain activation associated with an imaginary walk through hometown streets (Fig. 32), a mental imagery task they chose because people can perform it easily and have a vivid range of associations because of its familiarity. The illustration shows that this mental imagery activates areas in the back of the brain similar to those used in the primary visual experience—it is as if the subjects are actually seeing the landscape through which they are mentally walking.

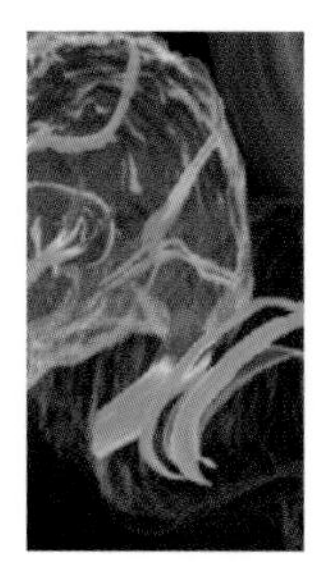

6. The Seventh Age of Man: Disease, Aging, and Death

BRAIN SICKNESS AND THE SEVEN AGES OF MAN

A melancholy awareness of the loss of powers with aging and the swiftness of the passing of time is noted more than once in Shakespeare's comedies. In As You Like It, *Jaques, a cynical lord attending the banished Duke Senior, speaks the play's most powerful lines. He tells us "All the world's a stage" and then, as a member of the cast on that broader stage (a cast that also, of course, includes the audience), he reminds us of the cruel brevity of our lives, which loop back from the dependence of infancy to dependence in the second childhood of senile old age. It is a theme that Shakespeare developed more fully in several plays, particularly and pointedly in* King Lear.

Jaques: All the world's a stage,
And all the men and women merely players.
They have their exits and their entrances,
And one man in his time plays many parts,

As You Like It. Floyd King as Jaques.

His acts being seven ages. At first the infant,
Mewling and puking in the nurse's arms.
Then, the whining school-boy with his satchel
And shining morning face, creeping like snail
Unwillingly to school. And then the lover,
Sighing like furnace, with a woeful ballad
Made to his mistress' eyebrow. Then, a soldier,
Full of strange oaths, and bearded like the pard,
Jealous in honour, sudden, and quick in quarrel,
Seeking the bubble reputation
Even in the cannon's mouth. And then, the justice,
In fair round belly, with good capon lin'd,
With eyes severe, and beard of formal cut,
Full of wise saws, and modern instances,
And so he plays his part. The sixth age shifts
Into the lean and slipper'd pantaloon,
With spectacles on nose, and pouch on side,
His youthful hose well sav'd, a world too wide
For his shrunk shank, and his big manly voice,
Turning again toward childish treble, pipes
And whistles in his sound. Last scene of all,
That ends this strange eventful history,
Is second childishness and mere oblivion,
Sans teeth, sans eyes, sans taste, sans everything.

As You Like It, 2.7

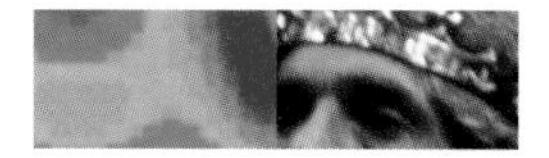

The Seventh Age of Man

Last scene of all, That ends this strange eventful history, Is second childishness and mere oblivion, Sans teeth, sans eyes, sans taste, sans everything.

The general fascination Shakespeare had for phenomena of the mind is illustrated further by his many references to the effects of diseases on the brain. He sometimes exploits the dramatic possibilities of distortions of the mind by disease to exaggerate specific features in his characters, such as the guilt of Lady Macbeth. In other instances, a point of plot turns on a disorder of the mind, such as Caesar's collapse before the people in an epileptic fit. In still others, Shakespeare uses disease to evoke pity for the once powerful. For no character is this more moving than for King Lear, when his beloved daughter

Cordelia's French troops find him after days of wandering. That scene presents the once great king as simply a "foolish, fond old man," as Lear describes himself. As we have followed him from the opening of the play, his decisions become less sound, his actions less rational, and his emotions more uncontrollable—in a pattern that calls to mind earlier stages in the pitiless development of Alzheimer's disease. Jaques is chronicling just such a life and its progression into a "second childishness and mere oblivion,/. . . sans everything."

Of course, Shakespeare would have known dementia and other diseases of the brain only through the changes in behavior that they cause. The first "modern" descriptions linking symptoms directly to changes in the brain were not to come until soon after Shakespeare's death in the publications of Thomas Willis (1621–1675). Now brain scientists are using imaging techniques to define the progression of brain diseases more precisely than is possible from trying to relate behavior in life to changes in the brain seen only after death (Fig. 33). The CT or MRI scan has become a standard part of the evaluation of patients with neurological problems. As shown in the images on the right in the illustration here, an MRI scan can crisply define the damage of a stroke. This makes it possible to rule out other causes of neurological symptoms (such as a brain tumor) and helps establish both the cause of the stroke and its prognosis.

Degeneration of the brain from Alzheimer's disease can be mapped by sensitive methods that allow comparisons of brain size and shape over time. This lends a remarkable aesthetics to the grim process of watching the brain die. For example, the images in the center and lower left from Paul Thompson's laboratory at the University of California at Los Angeles show how comparisons can be made between brains in attempts to distinguish the normal mild atrophy that accompanies aging from the severe, rapid degeneration of Alzheimer's disease.

Of course, brain disorders are not exclusively the province of the elderly. Young people can be stricken with diseases such as multiple sclerosis (MS), which produces variably distributed

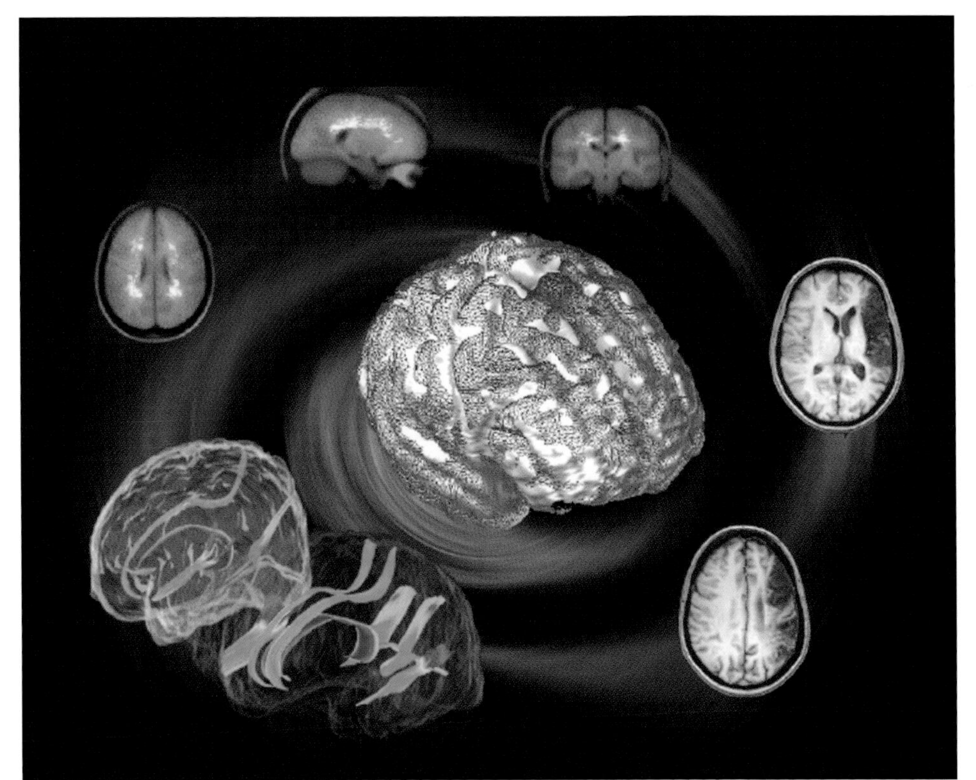

Figure 33. *Defining patterns of changes in brains with disease in the living patient is a critical first step for diagnosis and for monitoring the effects of treatment. Novel approaches to brain imaging and its analysis are providing new tools for doing this. The central image here shows a way of representing the average brain structure for a group of patients, and that on the lower left shows variations in the pattern of major folds in the brain across a population of Alzheimer's disease patients. The three images on the upper left show the distributions of lesions in the brains of patients with multiple sclerosis. Particular areas of the brain are most susceptible to changes from this disease. Conventional MRI images of individual subjects (as illustrated with the stroke patient images on the right-hand side) allow mapping of specific lesions with great anatomical detail.*

regions of inflammation in the deep (white) substance of the brain (where the myelin-coated axons that conduct signals between neurons are found). This is shown in the images in the top and upper left of the picture, which use a composite of brain scans from many patients to display the likelihood of lesions appearing in different parts of the brain with MS.

While there is always sadness in contemplating the causes of our mortality, we have to learn to understand diseases before we can change their courses or, ultimately, defeat them. Shakespeare used the dramatic potential of diseases of the brain to great effect, knowing that audiences can easily be moved by the strange powers of these disorders to alter the mind. Much of the power of Jaques's speech lies in the poignancy of joining the hopeful image of the new infant with the end stages of a demented old age "sans teeth, sans eyes, sans taste, sans everything." In *King Lear,* we are moved by the vulnerability of the king because it reflects our own vulnerability, or calls to mind an aging friend or relative. Now the new tools of brain imaging are providing hope to patients and their doctors by speeding accurate diagnosis, which allows appropriate therapies to be delivered more quickly. Advances in diagnostic and therapeutic techniques are also opening up new prospects for limiting some of the ravages of age and for reducing neurological disabilities. The specific markers of disease that brain imaging provides are helping scientists to precisely define drug actions in order to develop effective therapies more swiftly and at lower cost.

THE FALLING SICKNESS

Shakespeare clearly described several specific diseases affecting the brain. In Julius Caesar, *for example, he provided a brief but accurate description of epilepsy, which was originally called the falling sickness. (In fact, Shakespeare himself popularized the adjective* epileptic, *which is perhaps first used in written form in* King Lear.*) Casca, a Roman nobleman, tells Brutus, Cassius, and the other conspirators about a seizure that made Caesar fall down in the Roman marketplace just as a crowd of citizens was clamoring for him to become king. Discounting Caesar's apparent refusal to accept the crown, Cassius puns on "falling sickness" as the loss of political influence that the nobles of Rome face with Caesar's increasing power. The conspirators fear the growing adulation for Caesar among the common people.*

Cassius:	But soft, I pray you; what, did Caesar swound?
Casca:	He fell down in the market-place, and foam'd at mouth, and was speechless.
Brutus	'Tis very like; he hath the falling-sickness.
Cassius:	No, Caesar hath it not; but you, and I, And honest Casca, we have the falling-sickness.
Casca:	I know not what you mean by that, but I am sure Caesar fell down. If the tag-rag people did not clap him and hiss him, according as he pleas'd and displeas'd them, as they use to do the players in the theatre, I am no true man.
Brutus:	What said he when he came unto himself?

He fell down in the market-place, and foam'd at mouth, and was speechless.

Casca: Marry, before he fell down, when he perceiv'd the common herd was glad he refus'd the crown, he pluck'd me ope his doublet, and offer'd them his throat to cut. And I had been a man of any occupation, if I would not have taken him at a word, I would I might go to hell among the rogues. And so he fell. When he came to himself again, he said, if he had done or said anything amiss, he desir'd their worships to think it was his infirmity. Three or four wenches, where I stood, cried, 'Alas, good soul,' and forgave him with all their hearts; but there's no heed to be taken of them; if Caesar had stabb'd their mothers, they would have done no less.

Julius Caesar, 1.2

Epilepsy is among the most awe-inspiring of disorders. Sufferers typically experience sudden painless losses of consciousness, often with jerking of the limbs or twitching of the face. With some forms of epilepsy, the loss of consciousness is characterized simply by spells of staring. There may be a complete lack of awareness of the events and, most typically, no appreciation for the real time elapsed.

Loss of consciousness may be preceded by sensations of an "aura," a personal and sometimes complex experience. The aura may involve a visual hallucination (a patient of one of the authors described the sense of a "shadowlike" figure appearing just behind him during his aura, for example) or it can be highly charged emotionally (many sufferers describe an unreasoning sensation of fear). The aura may also be auditory. A patient of the famous Canadian neurosurgeon Wilder Penfield described to him how she heard the sound of a flock of birds rising into the sky before her seizures.

Epilepsy has been recognized since earliest times. As suggested by the wenches' cry ("Alas, good soul") and the wonder inspired by Caesar's collapse, some peoples have believed that those affected

Julius Caesar. John Nettles as Brutus and Julian Glover as Cassius.

by epilepsy have been given a special distinction, that the epileptic has been possessed briefly and "touched by the gods."

Seizures in epilepsy involve abnormal, synchronized firing of large groups of neurons in the cortex (the folded outer layer of the brain where neurons are found). Figure 34 illustrates a modern approach to localizing this abnormal neuronal activation non-invasively, using the technique of magnetoencephalography (MEG). When neurons are activated a small local electrical current is generated that has an associated magnetic field. As we briefly noted in the Introduction to this book, the highly sensitive detectors used for MEG can record the extraordinarily small magnetic fields generated by the brain in this way. If multiple detectors are arrayed around the skull, the individual regions of the brain that become activated can be localized. This allows the locations of these activated regions to be mapped, showing the propagation of the abnormal electrical activity away from the point of origin of the seizure, or "ictal focus" (a term derived from the Latin word *ictus,* meaning "stroke"). Each of the images has captured a different period of epileptic discharges, which consistently mapped to the same place in the brain. This gave doctors confidence that a single region was responsible for all of the patient's seizures.

In the computer-generated images here, the brain is represented in an "inflated" view, in which the surface convolutions have been flattened. The origin of the epileptic discharge can be seen at the frontal tip of the brain, as the signal is strongest here in the earliest moments of the seizure. Subsequent images demonstrate the progression of activation away from the "ictal focus," as the epileptic activity spreads farther through the brain. It is intriguing to observe that the spread of the waves is discontinuous. The discontinuity suggests that long fibers connecting widely spaced nerve cells in different brain regions are involved in the spread of the seizure, as well as short pathways connecting adjacent areas of cortex.

Seizures are generally treated in one of two ways. The most common treatment is the use of anticonvulsant agents, which reduce the tendency to develop a seizure, although they have no effect on the underlying cause of the seizures. A broad range of medicines now enables most people with epilepsy to achieve good control over their symptoms without suffering from unpleasant side effects.

For seizures resistant to these drugs, surgery is a useful and potentially curative alternative therapy. The key problem in epilepsy surgery is to identify the abnormal area of the brain as

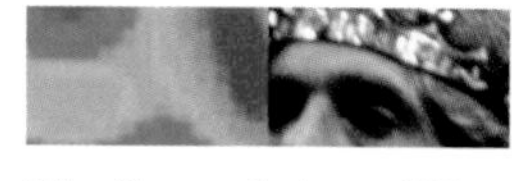

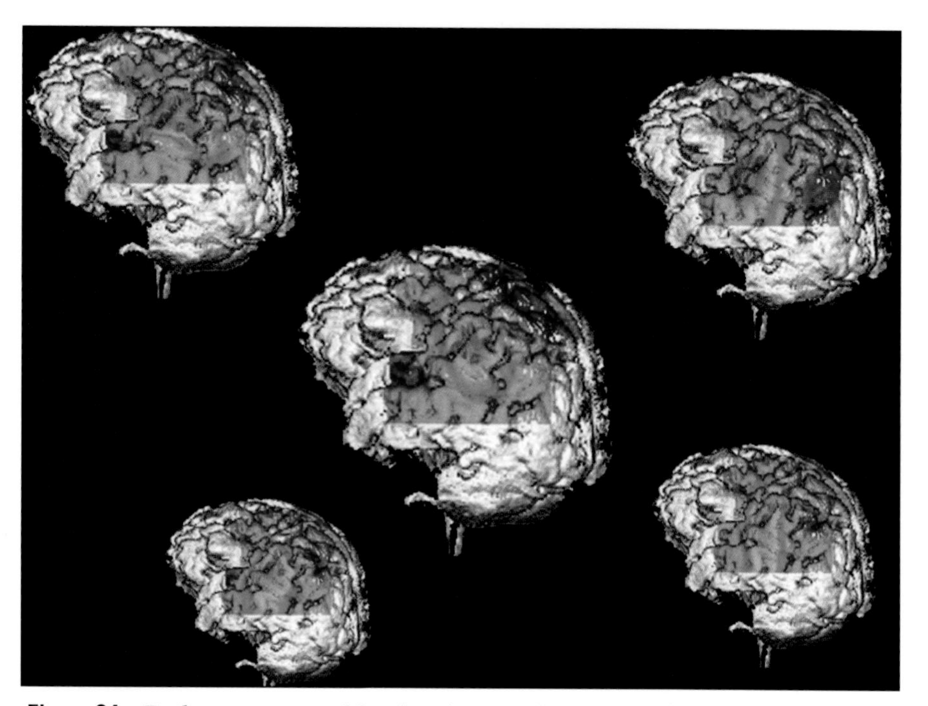

Figure 34. *Epilepsy is caused by the abnormal repetitive firing of large clusters of nerve cells. In focal epilepsies the abnormal firing can be traced to a specific region of the brain using tools such as electroencephalography, which detects the electrical signals directly, or magnetoencephalography (MEG), which measures the very small (on the order of 100,000,000 times smaller than the earth's magnetic field!) magnetic fields associated with these discharges. The images here illustrate the mapping of seizure discharges arising from an injured brain around a small cyst using MEG. With this information locating the "focus" of the patient's seizures, neurosurgeons were able to remove the abnormal brain tissue to cure the epilepsy.*

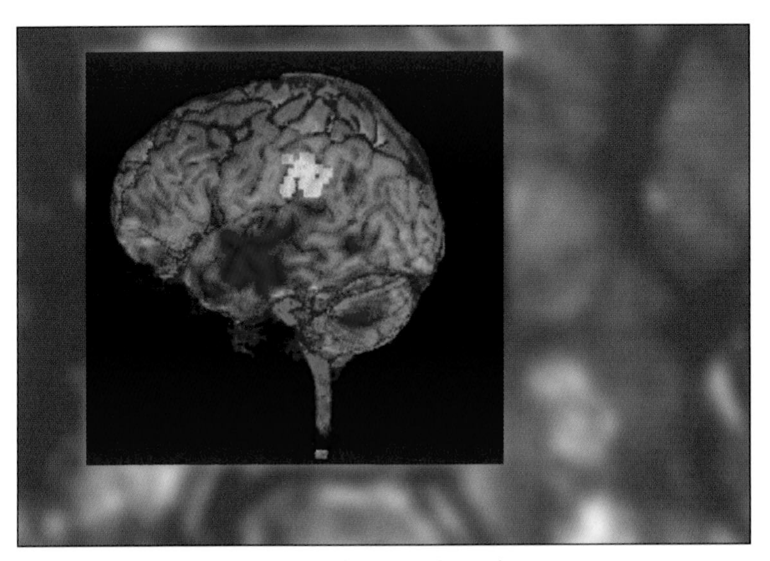

Figure 35. *Functional imaging techniques have become important because abnormal brain function typically cannot be seen from the surface, even if the brain can be observed directly. Here the cortex of a 19-year-old girl with seizures has been exposed at surgery, immediately over the region of brain responsible for her seizures. The seizure focus (white) was defined using the technique called single-photon-emission computed tomography (SPECT), which involved injection of a radiotracer compound immediately after the seizure. A radiotracer will be more concentrated in a brain region with increased activity from a seizure. The SPECT image then was merged with a three-dimensional reconstruction of the patient's brain MRI scan. This allows the neurosurgeon to understand the position of the abnormal tissue relative to the brain's surface anatomy. A major concern with this surgery in the left hemisphere was to avoid damaging brain areas used for language. MEG therefore was used to map (red) those regions of the brain that became active with speech. With this knowledge, the neurosurgeon was able to remove the seizure focus without injuring essential parts of the brain.*

precisely as possible. A first step is to employ noninvasive techniques such as electroencephalography (EEG, which directly detects the electrical waves in the brain) or the newer method of MEG (as shown here) to localize the area responsible for the abnormal electrical discharge, or the epileptic

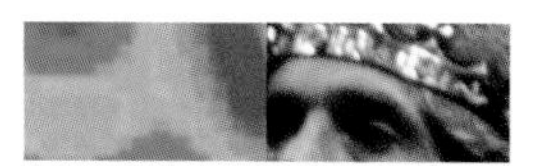

"focus." MRI scanning can be used to look for structural damage in the brain that might mark such a region. Abnormal brain metabolism associated with the seizure focus can be defined using techniques such as single-photon-emission tomography (SPECT; Fig. 35).

To ensure that removal of the affected part of the brain will not lead to new disability, sophisticated functional tests of brain activity sometimes must be performed. Invasive monitoring of the brain may be used in difficult cases in order to provide a more detailed map of the seizure activity and related normal brain functions. This is done by surgically implanting a grid of electrodes on the surface of the cortex for a brief period. Another strategy involves temporarily anesthetizing the part of the brain that would be affected by the surgery, to test the effects of the loss. However, as their reliability is being confirmed alternative, noninvasive approaches using brain imaging methods are increasingly being used. For example, MEG can define which parts of the brain are used for crucial language functions. This information then can be integrated with structural and metabolic information from MRI and SPECT, respectively, to show the neurosurgeon both where to find the lesion to be removed and those areas of normal brain that must be preserved.

A STRANGE COMMOTION IN THE BRAIN

Shakespeare's last history play, Henry VIII, *or* All Is True, *dramatizes the life of the great Tudor king and the birth of his daughter Elizabeth (who was, of course, the reigning queen in Shakespeare's time).* Henry VIII *includes a spectacular royal procession and is also remembered as the play during whose 1613 performance the Globe Theatre caught fire and burned to the ground. In the play, Cardinal Wolsey becomes moody because of his fears that Henry will learn of his great wealth and his efforts to block the king's divorce from Katherine of Aragon, his first wife. The Duke of Norfolk mentions the disturbed behavior of the cardinal to the king, who has Sir Thomas Lovell call Wolsey forward to explain his frenzied tics and starts. Although there is no evidence that Wolsey suffered from a movement disorder, Norfolk's description of Wolsey's curious behavior provides an opportunity to consider the symptoms of these remarkable disorders.*

Norfolk: My lord, we have
Stood here observing him. Some strange commotion
Is in his brain; he bites his lip, and starts,
Stops on a sudden, looks upon the ground,
Then lays his finger on his temple; straight
Springs out into fast gait, then stops again,
Strikes his breast hard, and anon he casts
His eye against the moon: in most strange postures
We have seen him set himself.

Henry VIII: It may well be,
There is a mutiny in's mind. This morning
Papers of state he sent me, to peruse
As I required: and wot you what I found
There (on my conscience put unwittingly)
Forsooth an inventory, thus importing
The several parcels of his plate, his treasure,

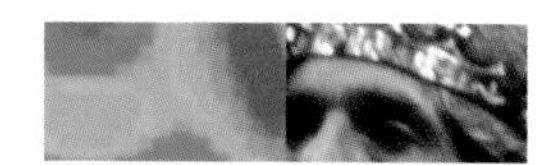

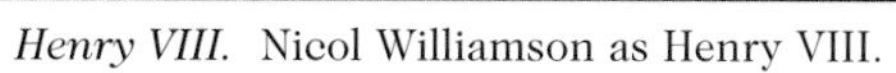

Henry VIII. Nicol Williamson as Henry VIII.

Rich stuffs and ornaments of household, which
I find at such proud rate, that it out-speaks
Possession of a subject.

Henry VIII, 3.2

Some strange commotion
Is in his brain;
he bites his lip,
and starts,
Stops on a
sudden, looks
upon the
ground,
Then lays his
finger on his
temple; straight
Springs out
into fast gait,
then stops
again....

Norfolk's description of a "strange commotion"—with biting of the lip, sudden starts and stops, rapid changes in gait, and strange postures—catalogues behaviors that we might see today as being associated with a number of brain diseases of the type known as movement disorders. Although with a different intention, the king also suggests that Wolsey's problem is a "mutiny in's mind."

If this were a movement disorder, what might Wolsey have been suffering from? The most common movement disorder (affecting 20 out of 100,000 people) is Parkinson's disease, which results from a progressive degeneration of the specialized brain neurons that release the chemical neurotransmitter, dopamine. Without adequate dopamine, movements become slow and stiff. Sufferers characteristically find it difficult to initiate movements. They may move forward with stops and starts and a characteristic shuffling, small-stepped gait. Norfolk's description brings this to mind as he describes how Wolsey "Springs out into fast gait, then stops again." Remarkably, sufferers can be jolted into action by, for example, a sense of emergency: a person with Parkinson's disease may generally move with painful slowness but still be able to leap up and run as an "automatic" response to a shout of "Fire!" Parkinson's disease sufferers typically also develop an obvious tremor at rest. Fortunately, the slowness of movement and the tremor respond well (at least in the earlier stages) to treatment with drugs such as L-dopa that enhance the disease-limited dopamine release in the brain.

Early in the twentieth century, an epidemic of the viral brain infection encephalitis lethargica (a "sleeping sickness") caused an unusual form of Parkinsonian syndrome. The neurologist Oliver Sacks described patients with this syndrome in his book

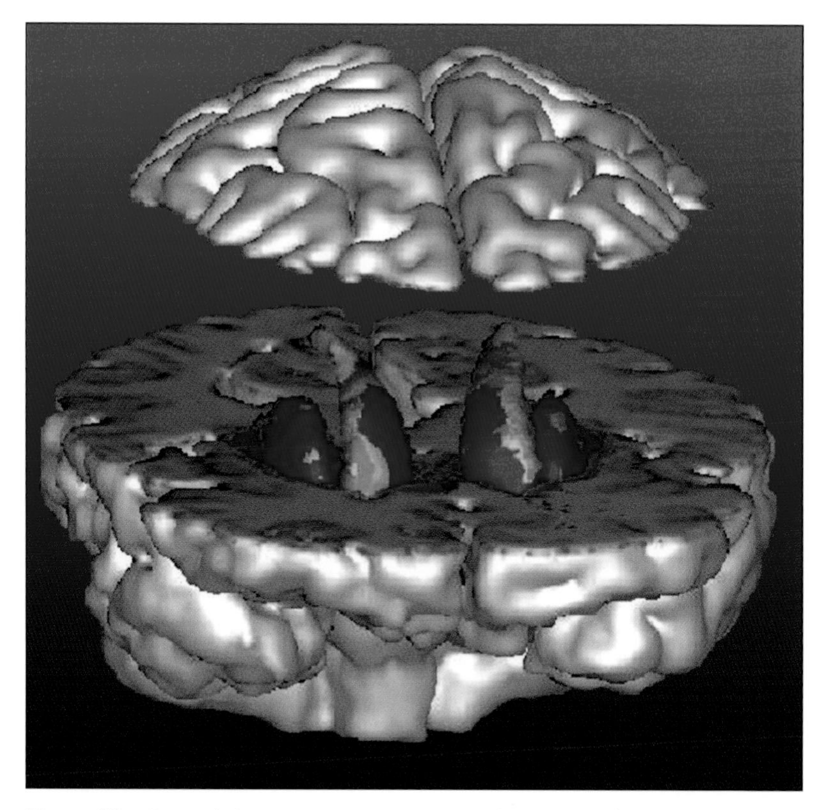

Figure 36. *One of the most important applications of PET brain imaging has been to define the distribution of the molecules on the surface of brain cells (called receptors) that are responsible for the actions of the signaling molecules (called neurotransmitters), which allow communication between nerves. The effects of different neurotransmitters are distinct and vary with the specific type of receptor with which they interact on the surface of a nerve cell. This image shows the distribution of receptors for the neurotransmitter dopamine. Selective death of nerve cells responsible for the production of dopamine occurs in Parkinson's disease, which impairs movements and causes a tremor of the hands. In this PET image, the very high density of receptors for dopamine is shown in a group of nerve cells deep in the brain, in the region known as the basal ganglia.*

Awakenings, which was made into a popular film starring Robin Williams and Robert DeNiro. A remarkable part of Sacks's story was his description of the effects of L-dopa in this rare disease. In a fashion generally more exaggerated than is found in typical Parkinson's disease, these postencephalitic Parkinson's syndrome patients showed sudden, dramatic changes in their movements after receiving the drug. Some developed hyperactivity, with sudden flinging of their limbs. These movements were uncontrollable and would erupt explosively and then end just as quickly in stillness. Rather as Cardinal Wolsey "casts/His eye against the moon," some of the patients had frightening "oculogyric crises," in which their eyes rolled upward uncontrollably. The side effects were so great in many of the patients described by Sacks that although the treatment allowed them to move from their previously disease-frozen, stiff postures, they were unable to continue with it. Fortunately, such severe side effects are not typical when L-dopa is given to patients with the more usual form of Parkinson's disease.

A more tragic movement disorder is the genetically transmitted Huntington's disease. Huntington's disease results from the inheritance of a mutant gene that causes formation of an abnormally long form of a normal brain protein (now called huntingtin). Although the usual function of this protein is still not understood, it is clear that the protein is toxic to nerve cells. The result is that as individuals carrying the genetic defect age, cells in the brain progressively die. These changes occur early in a key movement center of the brain, the basal ganglia, located deep in the white matter of the brain. In addition to changes in personality and thought, sufferers may experience constant and terrifyingly bizarre movements. They may develop repetitive tics with uncontrollable grimacings of the face. Their limbs may be flung about violently and purposelessly. At the same time, they may develop progressively more severe cognitive impairments and eventually die. It is terrible to consider that death from the most severe manifestations of disease may come as a relief to both sufferers and their families.

A much less worrying disorder associated with abnormal movements is Tourette's syndrome, which is associated with repetitive tics and often frequent expletives. Many individuals with Tourette's syndrome have "coprolalia"—a compulsion to swear. Tourette's is a nonprogressive disorder and is compatible with a perhaps extraordinary but perfectly fulfilling life. Many people who have this disorder are never even diagnosed as having a disease.

Figure 36 illustrates the way in which noninvasive PET scanning can be used to study the causes of these and related disorders at a molecular level. Subjects scanned to make this type of image are given a very tiny dose of a chemical that binds strongly to the part of the surface of the nerve cell that interacts directly with the signaling molecule dopamine. Here the bright colors define the distribution of these dopamine receptors in the basal ganglia, the deep brain area that is involved with all three of the movement disorders discussed. This powerful technology is helping to drive the development of new drugs for the treatment of these disorders and an improved understanding of why currently available drugs begin to fail as disorders such as Parkinson's and Huntington's progress.

THE BLACKNESS OF DEPRESSION

Probably the most poetic of Shakespeare's history plays, Richard II *presents a monarch more in control of his language than his country. In the course of the play, King Richard comes to recognize his weakness not only as a ruler but also as a man. His banished cousin, Bolingbroke, returns to depose him, and Richard learns from his followers Scroope and Aumerle that many of his supporters have been beheaded. Forlorn, he delivers a stirring eulogy for dead kings, whose greatness is so ephemeral that it can be undone by seemingly little more than the prick of a "little pin." There are few instances in Shakespeare's plays in which a character displays the blackness of depression more clearly than does Richard in this scene.*

Richard II: No matter where—of comfort no man speak.
Let's talk of graves, of worms, and epitaphs,
Make dust our paper, and with rainy eyes
Write sorrow on the bosom of the earth.
Let's choose executors and talk of wills.
And yet not so—for what can we bequeath
Save our deposed bodies to the ground?
Our lands, our lives, and all, are Bolingbroke's,
And nothing can we call our own but death;
And that small model of the barren earth
Which serves as paste and cover to our bones.
For God's sake let us sit upon the ground
And tell sad stories of the death of kings:
How some have been depos'd, some slain in war,
Some haunted by the ghosts they have deposed,
Some poisoned by their wives, some sleeping kill'd,
All murthered—for within the hollow crown
That rounds the mortal temples of a king
Keeps Death his court and there the antic sits,
Scoffing his state and grinning at his pomp,

Richard II. Jeremy Irons as Richard II.

Allowing him a breath, a little scene,
To monarchize, be fear'd, and kill with looks;
Infusing him with self and vain conceit,
As if this flesh which walls about our life
Were brass impregnable; and, humour'd thus,
Comes at the last, and with a little pin
Bores through his castle wall, and farewell king!
Cover your heads, and mock not flesh and blood
With solemn reverence; throw away respect,
Tradition, form, and ceremonious duty;
For you have but mistook me all this while.
I live with bread like you, feel want,
Taste grief, need friends—subjected thus,
How can you say to me, I am a king?

Richard II, 3.2

Let's talk of graves, of worms, and epitaphs, Make dust our paper, and with rainy eyes Write sorrow on the bosom of the earth.

Depression and its consequences are estimated to cost Americans more each year than cancer, coronary heart disease, and AIDS combined. It is a major health concern now and must have been similarly common in Shakespeare's time. Shakespeare returned to the problem again and again in different ways with many characters in his plays, but few examples are as powerful as that of Richard II as he contemplates his future.

Depression can be triggered by specific events (as in Richard's case) or it may occur spontaneously. Shakespeare here provides us with an almost perfect description of many of the clinical features of depression. Like Richard, a depressed patient can show low mood, a lack of normal interest and enjoyment in the world, and an abnormal, increased sense of tiredness. Features such as a loss of self-worth, feelings of guilt or unworthiness, and a very bleak and pessimistic view of the future are characteristic. Patients also have other symptoms including unexplained tearfulness and disorders of sleep.

Richard tells us immediately of his low mood as he whispers, "of comfort no man speak./Let's talk of graves, of worms, and epitaphs." His joy in life has gone, and he feels able only to

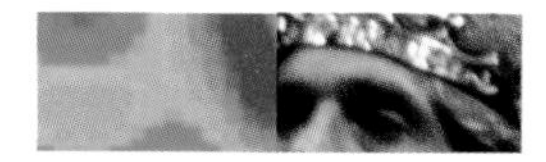

"tell sad stories of the death of kings." His speech drags with pessimism as he recalls other monarchs—"all murthered." He is waiting for death's "little pin" that will bring his end, signaling "farewell king!" He no longer sees himself as one chosen by the hand of God to rise above his fellow men, but merely as a weak man "within the hollow crown/That rounds the mortal temples of a king."

Shakespeare appreciated how depression can incapacitate a sufferer. He also saw the risk of suicide, which marks depression as a potentially fatal disease. The next chapter of this book considers the madness of Ophelia as she retreats into dark, psychotic despair over the loss of Hamlet's love. One of the preventable tragedies of depression is that it is too often not recognized and treated early enough. It can be far advanced before the depressed person, or those close to him or her, even recognize that it has taken hold.

A major issue in the care of depressed patients is to distinguish those who will benefit rapidly from therapy (and can therefore be treated as outpatients) from those who require special attention or hospitalization for their own protection. Helen Mayberg of the University of Toronto (then working at the University of Texas at San Antonio) has made observations which suggest that PET (or related functional brain imaging techniques) may one day allow physicians and other caregivers to distinguish good treatment responders from others. Mayberg's group contrasted the patterns of brain metabolism in a group of patients who were depressed but had responded well to treatment with those of a group that had not. They found a difference between the groups in a small area near the middle front of the brain called the anterior cingulate cortex. This part of the brain has strong connections to other areas that make up the so-called limbic, or emotional, system of the brain. Patients who responded well to treatment showed increased blood flow in this region, suggesting that their nerve cells were more active, compared with those who did not respond to treatment or with normal control subjects. This

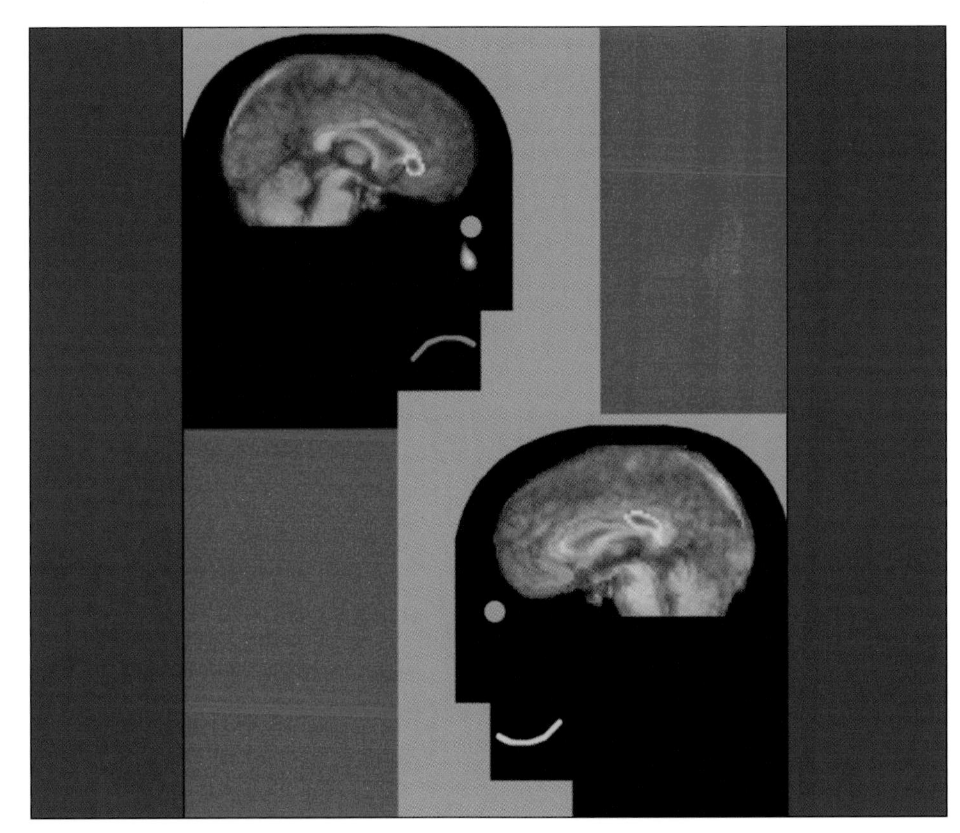

Figure 37. *A fundamental goal of psychiatry has been to understand the specific ways in which the brain becomes dysfunctional in the major psychiatric syndromes. Sadness and depression have long been recognized to be distinct. This figure shows a composite of PET images of brain metabolism in subjects who are depressed (upper left) and in those who have recovered from depression (lower right). There are distinct differences in the patterns of brain activation. Depressed patients show greater activity (yellow-red areas) in the more anterior part of a deep brain structure known as the anterior cingulate, while recovered subjects show greater activity posteriorly. There thus appears to be a reciprocal relationship between these two areas of the cortex, with increases in activity in one being associated with decreases in activity in the other. These results suggest that a characteristic of depression may be a reversal of the usual relationship between these areas.*

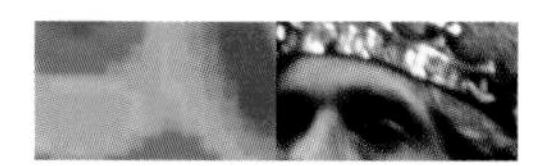

suggests that the anterior cingulate cortex may play a central role in normalizing the functional abnormalities that accompany depression. Perhaps differences in the way this part of the brain works distinguish those who have only brief periods of unhappiness from those who develop the full clinical manifestations of depression.

Figure 37 illustrates the results of another experiment performed by Mayberg and her colleagues, in which patients were studied while depressed and then after recovery. A more frontal region of the anterior cingulate cortex showed increased activity during the period of depression (indicated by the bright red in the upper image). With recovery, the abnormally increased metabolism seen on the PET scan decreased to normal levels. Reciprocal changes were seen further back in the brain. These results suggest that abnormal activity in rather discrete areas of the brain is associated with (and perhaps could cause) the symptoms of depression. In the future, use of PET and related functional imaging techniques may lead to more targeted treatments for sufferers of this terrible illness.

TEMPESTS IN THE MIND

King Lear *chronicles the dissolution of a kingdom and the death of a king. The storm that envelops the stage in Act 3 provides a powerful metaphor for the tempest stirring within Lear's mind. Lear's mounting rage at what he perceives to be his children's ingratitude ("I am a man/More sinned against than sinning") drives him to rush from the castle into a great storm with his loyal court jester and the Earl of Kent. However unpleasant, the raging storm is less awful to him than are the indignities heaped on him in his daughters' courts. As he goes into the wild weather, he tries to get his two followers to seek shelter from the storm while taking on the greater challenge of finding relief from his inner storms. Recognizing the terrible folly of having been deceived by his false daughters, Regan and Goneril, and of having punished the loyal Cordelia, he fears that he may be going mad.*

King Lear: Thou think'st 'tis much that this contentious storm
Invades us to the skin: so 'tis to thee,
But where the greater malady is fixed,
The lesser is scarce felt. Thou'dst shun a bear,
But if thy flight lay toward the roaring sea,
Thou'dst meet the bear i'the mouth. When the mind's free,
The body's delicate: this tempest in my mind
Doth from my senses take all feeling else,
Save what beats there, filial ingratitude.
Is it not as this mouth should tear this hand
For lifting food to't? But I will punish home;
No, I will weep no more. In such a night
To shut me out? Pour on, I will endure.
In such a night as this? O, Regan, Goneril,

King Lear. Anthony Hopkins as Lear.

Your old, kind father, whose frank heart gave you all—
O, that way madness lies, let me shun that;
No more of that.

King Lear, 3.4

When the mind's free, The body's delicate: this tempest in my mind....

Madness was a theme that Shakespeare returned to many times in his plays, and particularly vividly in *King Lear.* The king's fear that he is losing his mind arises from a disorder of thought that is part of his rage and thus eventually passes. Others, however, are not so fortunate. For example, patients with schizophrenia, a common psychiatric disease (affecting 1 percent of the population in cultures across the world), have disordered thoughts that persist and worsen. The symptoms of schizophrenia must have been known to Shakespeare, although it was not until the last century that the disease was specifically defined.

Auditory hallucinations are a key feature of schizophrenia. These hallucinations typically provide a running commentary to schizophrenia sufferers on their own actions. Such a "tempest in my mind" can make people suffering from schizophrenia unable to distinguish an internal voice from a real voice external to themselves. They may even believe that they hear two or more distinct voices arising from their own disordered thoughts. (Note that this is very different from the normal behavior of literally talking to oneself.)

Like Lear wandering in the stormy night, absorbed in the world of his own thoughts, patients with schizophrenia make up a tragically large part of our homeless population today. The most difficult aspect of the disease to treat (and thus the most challenging aspect) is the social isolation it inflicts on sufferers. In much the same way that Lear refrained from seeking immediate solace from Cordelia because of shame (and perhaps a fear of rejection), patients with schizophrenia may fear or distrust others and shy away from their help.

Schizophrenia is associated with impairment of functions performed by the frontal lobes. Functional brain imaging studies have

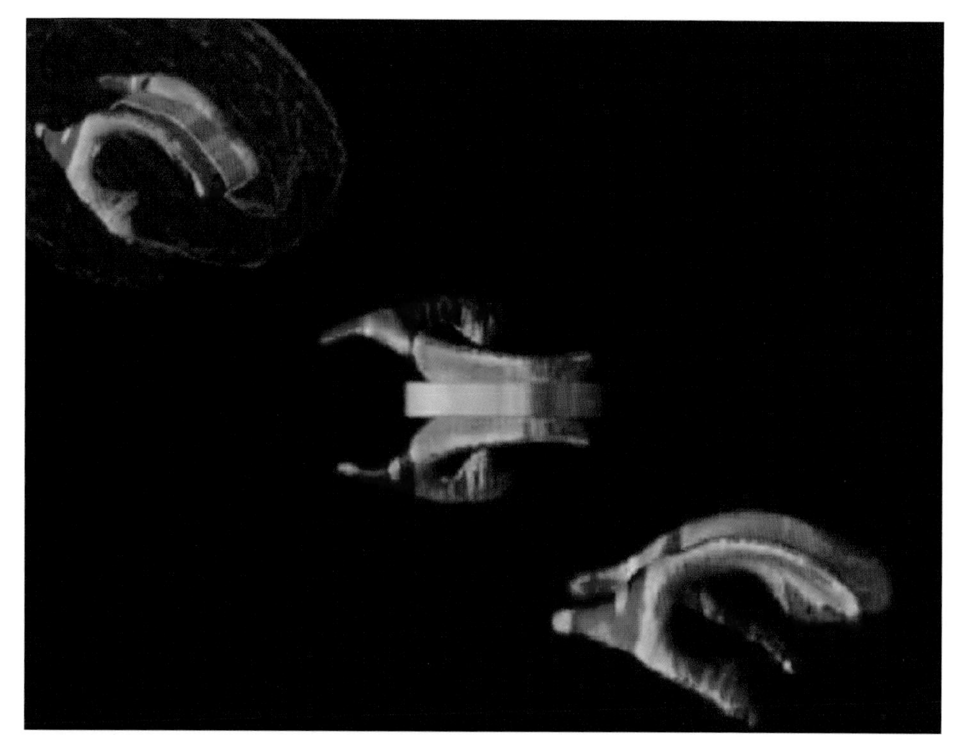

Figure 38. *Structural differences can be identified that distinguish the brain of a schizophrenic patient from that of a healthy person. To make these images, high-resolution MRI images were acquired from a group of schizophrenic patients and a group of healthy people of the same ages. The fluid-filled cavities in the center of the brain known as the ventricles then were outlined on each subject's scan. An "average ventricle" image then was constructed for the healthy subjects and for the schizophrenic patients. Here the magnitudes of differences between the two average maps are shown as a color scale (red indicating the largest differences; blue-purple indicating the smallest differences) on a model ventricular surface. The red areas confirm that ventricular enlargement, which occurs with brain atrophy (shrinkage), is a feature of the brain of a schizophrenic patient. However, the variation of color tells us that these changes are local to specific brain regions. While the nature of the brain systems cannot be defined directly using this method, results like these suggest that particular systems are affected primarily by the disease.*

shown, for example, decreased blood flow in the frontal cortex of schizophrenics. Structural changes also occur. Localized regions of brain atrophy can be found that suggest maldevelopment, injury, or degeneration occur in specific brain systems (Fig. 38).

Research at the Institute of Psychiatry in London has focused on trying to define the basis of the auditory hallucinations of schizophrenia. In an important study, researchers tested patients who were able to report precisely when their inner voices began to speak to them. The patients were asked to listen to a recorded story while their brains were being scanned by fMRI. When auditory hallucinations intruded on the experience, they signaled this by pressing a button. The imaging experiment measured the relative increase in brain activity during story reading relative to that during a period when the story was not played. Normally this should have led to clear changes in auditory and language centers in the brain.

It was found, however, that the extent of new brain activation in these areas, which normally become active when hearing speech, was substantially reduced in patients who were actively hallucinating. The researchers interpreted this as evidence that auditory hallucinations activate the same areas of the brain as normal hearing (and therefore interfere with activation of the same areas by real speech). It supports the idea that a key difference between the schizophrenic and the normal brain is the inability to sufficiently distinguish stimuli generated by the self from those generated externally. Fortunately, treatments for schizophrenia have been rather successful in relieving sufferers of their hallucinations.

7.  Drugs and the Brain

A CELEBRATION OF ALCOHOL

In various plays, Shakespeare both censures and celebrates the effects of alcohol on the human brain and body. For its celebration, he speaks through Falstaff, one of his most memorable characters and a man resolutely committed to drinking. In Henry IV, Part II, *Falstaff recounts the benefits of alcohol. Left alone onstage by the sober Prince John, Falstaff comically considers that some wine would help mellow the too-serious prince. As this prose monologue continues, he singles out Prince Henry (or Harry), John's older brother and the heir to the throne, as an example of one who has improved as a result of the effects of alcohol (a popular form of which at the time was "sherris-sack," a type of fortified wine similar to modern sherry).*

Falstaff: Good faith, this same young sober-blooded
boy doth not love me, nor a man cannot make
him laugh; but that's no marvel, he drinks no wine.
There's never none of these demure boys come to any
proof; for thin drink doth so over-cool their blood,

and making many fish meals, that they fall into a
kind of male green-sickness; and then when they
marry they get wenches. They are generally fools
and cowards—which some of us should be too, but for
inflammation. A good sherris-sack hath a twofold
operation in it. It ascends me into the brain,
dries me there all the foolish and dull and crudy
vapours which environ it, makes it apprehensive,
quick, forgetive, full of nimble, fiery, and
delectable shapes, which delivered o'er to the
voice, the tongue, which is the birth, becomes
excellent wit. The second property of your
excellent sherris is the warming of the blood,
which before, cold and settled, left the liver
white and pale, which is the badge of pusillanimity
and cowardice; but the sherris warms it and makes
it course from the inwards to the parts, extremes.
It illumineth the face, which, as a beacon, gives
warning to all the rest of this little kingdom,
man, to arm; and then the vital commoners, and
inland petty spirits, muster me all to their captain,
the heart; who, great and puffed up with this
retinue, doth any deed of courage; and this valour
comes of sherris. So that skill in the weapon is
nothing without sack, for that sets it a-work, and
learning a mere hoard of gold kept by a devil, till
sack commences it and sets it in act and use.
Hereof comes it that Prince Harry is valiant; for
the cold blood he did naturally inherit of his
father he hath like lean, sterile, and bare land
manured, husbanded, and tilled, with excellent
endeavour of drinking good and good store of fertile
sherris, that he is become very hot and valiant. If
I had a thousand sons, the first human principle I
would teach them should be to forswear thin
potations, and to addict themselves to sack.

Henry IV, Part II, 4.3

Henry IV, Part II. David Sabin as Falstaff and Derek Smith as Hal.

Good faith, this same young sober-blooded boy doth not love me, nor a man cannot make him laugh; but that's no marvel, he drinks no wine.

Falstaff is a rogue, a drunkard, and a coward. He is also the comic "life of the party" and deeply fond of his companions-in-carousing. But despite his rich good humor, one sees in him the tragic self-awareness of a life unfulfilled. His melancholy stems in part from recollections of slights, failures, and unmet expectations.

Falstaff has become enamored of the personality that he has generated for himself with the help of alcohol. His celebration of alcohol springs from the way it relieves self-doubt and banishes unpleasant memories. He needs to be freed from inhibitions arising from his own sense of unworthiness in order to enjoy himself. His celebration of alcohol is also interesting from a medical point of view because it both describes much that is true about this most common of drugs and includes myths that only a person "under the influence" (such as Falstaff) could propagate.

In trying to rationalize Prince John's dull seriousness, Falstaff concludes that part of the problem is that John simply "drinks no wine." Falstaff describes the effects of alcohol in reducing inhibition, claiming that "It ascends me into the brain, [and] dries me there all the foolish and dull and crudy vapours," allowing his "excellent wit" to surface.

Alcohol of course has more general effects on the body. The "warming of the blood" that Falstaff refers to is not real but a feeling that comes from increased circulation due to alcohol-induced dilation of the small vessels of the skin. In consequence, it "illumineth the face, which, as a beacon, gives warning to all the rest of this little kingdom, man . . . "—the "warning" likely being simply that the drinker has had too much!

Falstaff describes alcohol from the standpoint of an unrepentant abuser of the drug. In fact, alcohol is a depressant of the nervous system. It makes speech appear "nimble" because it depresses brain functions responsible for self-monitoring and other processes that demand focused attention. This loss of effective self-monitoring underlies the partygoer's insistence that he can drive home safely, when everyone else can see that he is too drunk even to walk straight. As many a would-be, drink-befuddled Casanova has discovered, too much alcohol can impair sexual function. As the castle gatekeeper jokes in *Macbeth* (2.3), drink "provokes the desire, but it takes away the performance."

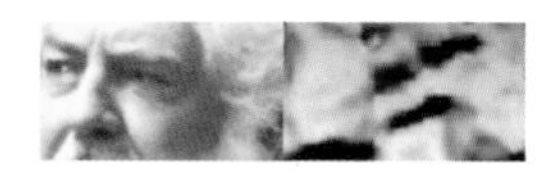

The remarkable extent to which alcohol can depress brain responses is illustrated by the fMRI scans shown here (Fig. 39), produced in a study of the mechanisms by which alcohol slows reaction times. A simple test of reaction time is to ask a person

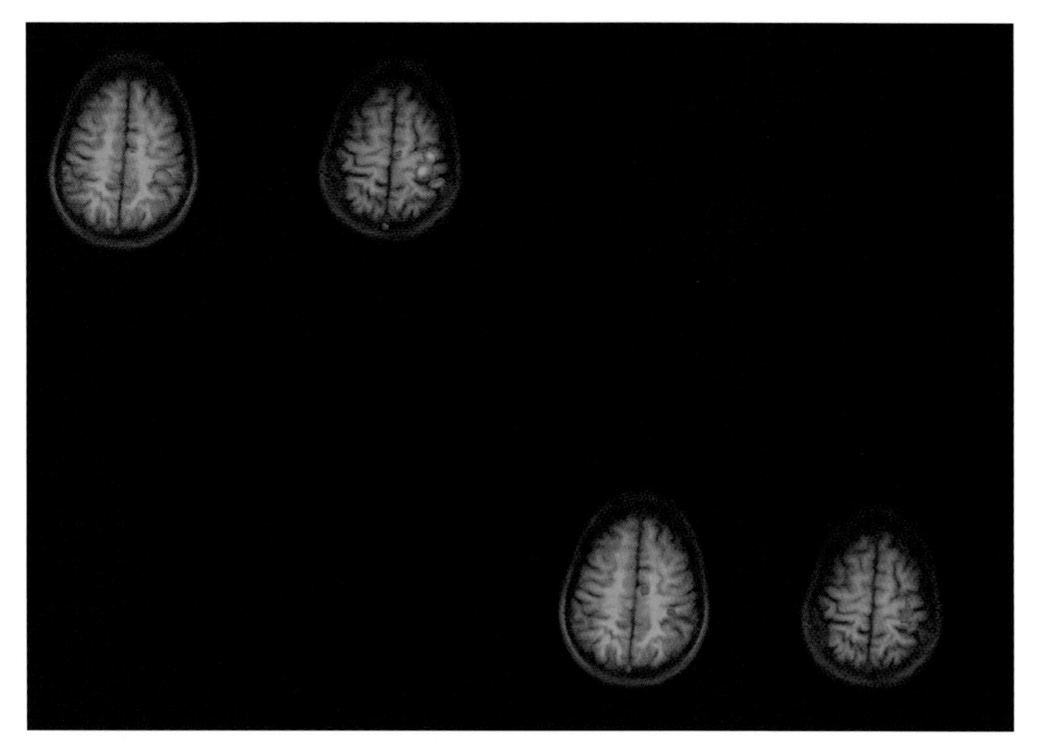

Figure 39. *Alcohol has a broad range of effects on the brain. To generate these fMRI scans, subjects were asked to push a button as quickly as possible when a right-facing arrow was presented on the screen in front of them. Brain activation in both motor planning (in the middle of the brain) and execution (to the side of the brain) areas was strong before alcohol was consumed (upper left). After a subject drank the equivalent of two cocktails over the space of a few minutes, the speed with which he or she could accurately push the button was slowed. In this experiment, there also appeared to be an accompanying decrease of all the brain responses being mapped (lower images).*

to press a button as quickly as possible after a simple cue is presented. It can be made more challenging by also presenting a conflicting cue. In the brain, this task involves rapid planning for a button press every time the action cue is presented, and inhibition of this action when the alternate cue appears. The normal brain response to such a "cued" movement involves prominent increases in activity of the motor planning region of the supplementary motor cortex and, more laterally, in an area known as the premotor cortex, as well as in the primary motor cortex in the hemisphere opposite the hand being moved. When a person ingests alcohol (in this experiment the equivalent of a double vodka drunk quickly), the reaction time for this response increases by about 50 percent. Associated with this is a dramatic diffuse reduction of the magnitude of brain activation detected by fMRI.

In *Henry V,* Shakespeare shows Falstaff to be a poor swordsman, so it is with tongue in cheek that he has Falstaff claim, "skill in the weapon is nothing without sack." However, while "valour comes of sherris"—or at least the valor that arises from not appreciating consequences—skill does not. As a visit to the bar after a big game can show, too much alcohol can make the finest athlete as clumsy as a toddler. Even moderate amounts of alcohol can impair motor reaction times enough to preclude safe driving.

MIMING DEATH

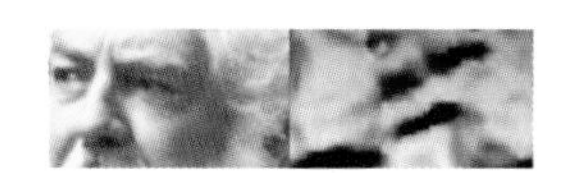

More than one of Shakespeare's plots rely on the use of sleep-inducing drugs or poisons. At the heart of Romeo and Juliet *lies a plan formulated by Friar Laurence to help the young lovers safely escape their warring families. Romeo has been banished from Verona for killing Tybalt, Juliet's cousin. To help reunite the couple, Friar Laurence gives Juliet a potion to drink that will cause her to appear dead, because he knows that her family will then place her in a burial vault outside the city walls. Friar Laurence's plan is that Romeo will soon afterward rescue her from the tomb. In this scene Juliet bids goodnight to her mother and nurse before swallowing the drug. In taking the drug, she must put all her trust in Friar Laurence, because she cannot be sure that the powerful potion will not really kill her.*

Juliet: Farewell. God knows when we shall meet again.
I have a faint cold fear thrills through my veins
That almost freezes up the heat of life.
I'll call them back again to comfort me.
—Nurse!—What should she do here?
My dismal scene I needs must act alone.
Come, vial.
What if this mixture do not work at all?
Shall I be married then tomorrow morning?
No! no! This shall forbid it. Lie thou there.

[She lays down a knife.]

What if it be a poison which the Friar
Subtly hath minister'd to have me dead,
Lest in this marriage he should be dishonour'd,
Because he married me before to Romeo?
I fear it is. And yet methinks it should not,
For he hath still been tried a holy man.
How if, when I am laid into the tomb,

Romeo and Juliet. Emery Battis as Capulet, Fran Dorn as Lady Capulet, and Laura Hicks as Juliet.

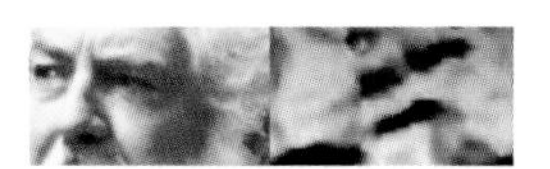

I wake before the time that Romeo
Come to redeem me? There's a fearful point!
Shall I not then be stifled in the vault,
To whose foul mouth no healthsome air breathes in,
And there die strangled ere my Romeo comes?
Or, if I live, is it not very like,
The horrible conceit of death and night
Together with the terror of the place,
As in a vault, an ancient receptacle
Where for this many hundred years the bones
Of all my buried ancestors are pack'd,
Where bloody Tybalt yet but green in earth
Lies festering in his shroud; where, as they say,
At some hours in the night spirits resort—
Alack, alack! Is it not like that I
So early waking, what with loathsome smells,
And shrieks like mandrakes torn out of the earth,
That living mortals, hearing them, run mad—
O, if I wake, shall I not be distraught,
Environed with all these hideous fears,
And madly play with my forefathers' joints,
And pluck the mangled Tybalt from his shroud,
And, in this rage, with some great kinsman's bone
As with a club dash out my desperate brains?
O look, methinks I see my cousin's ghost
Seeking out Romeo that did spit his body
Upon a rapier's point! Stay, Tybalt, stay!
Romeo, Romeo, Romeo, here's drink. I drink to thee!
[She falls upon her bed within the curtains.]

Romeo and Juliet, 4.3

I have a faint cold fear thrills through my veins That almost freezes up the heat of life.

We all can sympathize with Juliet's fears. Losing conscious control over the body is a frightening prospect. Preparing for anesthesia before surgery, for example, who among us would not feel at least something of the "faint cold fear . . . /That almost freezes up the heat of life"? When we are unconscious, we are unable to protect our bodies even in simple ways that we generally take for granted, such as by continuing to breathe or protecting our lungs from swallowed saliva.

To mime death, Juliet must not only lose her awareness but also forfeit her responsiveness to external stimuli such as sound and touch. She would also experience a dramatic slowing of heart rate and respiration. In fact, slowing respiration would have been particularly important to the success of Friar Laurence's plan, because a way Juliet's feigned death might have been exposed would have been for someone to hold a mirror beneath her nose to see if her breath caused it to fog.

What drug might Friar Laurence have given Juliet? A drug that induces some of the appearances of death when given in high doses is phenobarbital. Phenobarbital is a barbiturate, a type of drug that inhibits brain activity directly. Patients who have suffered severe brain injury have sometimes been given phenobarbital at very high doses to induce a so-called barbiturate coma—a state so close to death that the patients' vital functions must be maintained by life support systems. The main purpose of this risky procedure is to reduce the metabolic demands of the injured brain. Juliet appears instinctively to recognize the reduced need for oxygen in such a coma when she reflects that only if she awakes early will she run the risk of being "stifled in the vault,/To whose foul mouth no healthsome air breathes in."

However, barbiturates were not available in Shakespeare's time. More likely, Friar Laurence gave Juliet an opiate. These drugs, derived from the opium poppy, include codeine, morphine, and heroin. They are most important medically for pain relief. They also have more general effects on the brain, initially inducing a sense of well-being (which accounts in part for their addictive potential), and, at higher doses, depressing consciousness and respiration.

Choosing the right dose for these drugs is critical. Pain must be relieved adequately but not at the expense of blood pressure or respiration, unless these can be controlled by other methods. In a modern operating room, anesthetists monitor the vital signs of patients with extreme care on a minute-by-minute basis to minimize the hazards of these powerful agents. Juliet's concerns about the dose in her own phial are therefore not at all misplaced. At the wrong dose, even the right drug can be a poison—as Juliet worries, "What if this mixture do not work at all? . . . What if it be a poison . . . ?"

Defining the correct dosage for drugs that alter consciousness or pain perception would be greatly aided by objective

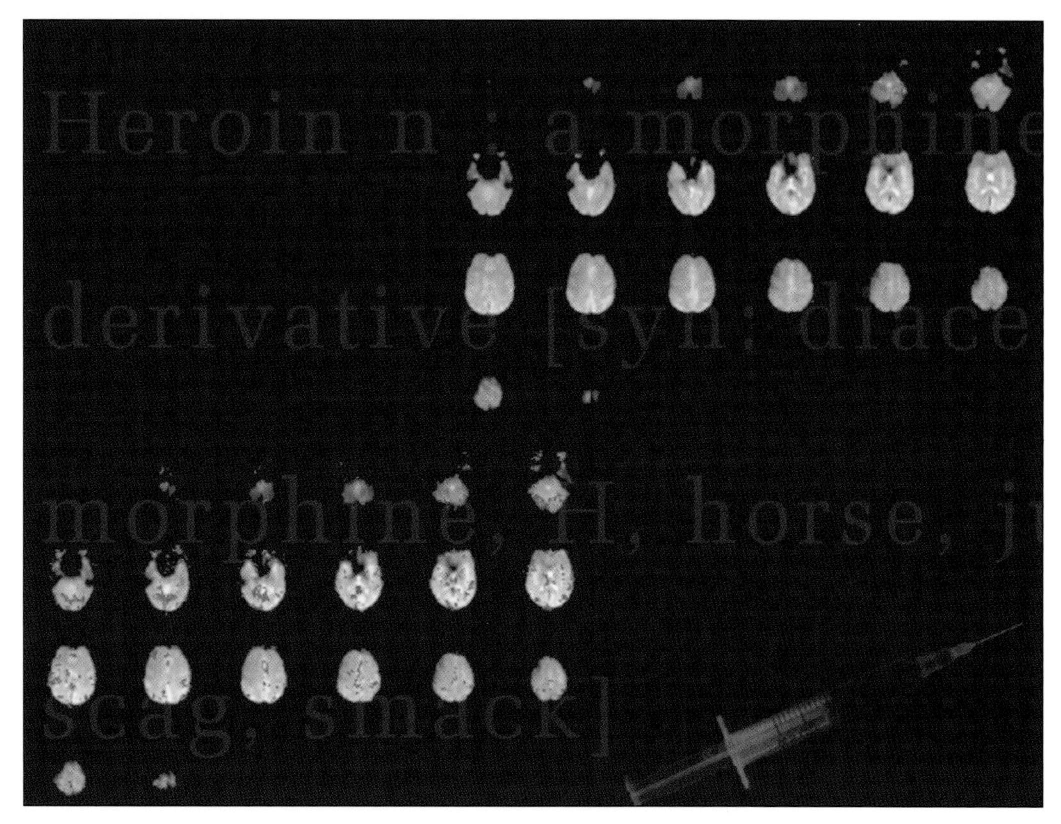

Figure 40. *Opiates are among the most powerful analgesic (pain-relieving) drugs known. They can also cause a feeling of drowsy euphoria and have been drugs of abuse for centuries. The fMRI images in the left panel show the complex pattern of brain activations that occur when a painfully hot piece of metal is applied to the back of the left hand. However, after infusion of an opiate (remifentanil), the same stimulation causes very little brain activation (right panel). A sense of well-being and a lack of perception of the painfulness of the hot stimulus were associated with the decrease in brain activation.*

measures of response. Recent work with fMRI by Irene Tracey's group in Oxford suggests a novel way in which this might be done (Fig. 40). In initial studies, they showed that a painful heat stimulus applied to the hand activates areas in the brain that respond to simple sensations as well as areas in the limbic system that are associated with the emotional response to the unpleasantness of the stimulus. The lower images in the illustration show where the brain becomes more active with the painful stimulus alone. When a drug called remifentanil (a very short-acting opiate) is given, the magnitude of response to the pain is reduced according to the dosage. At higher doses, almost no pain is felt, and the entire pattern of pain response in the brain is almost eliminated, as shown by the images in the upper right. This clearly demonstrates the opiate's ability to limit the brain's response to pain and other external stimuli—just what Friar Laurence would want for Juliet.

THE SEDUCTION OF DRUGS

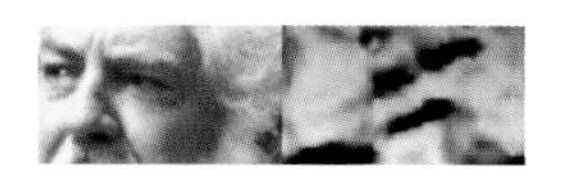

The power of drugs to impair judgment or weaken the will is referred to in several places in Shakespeare's plays. A particularly clear example occurs in the tragedy of Othello. *When Othello secretly marries the fair-skinned Desdemona, her father, Brabantio, is unable to understand how his daughter could have chosen to wed the Moor. Brabantio accuses Othello of using sorcery to trap her "with drugs or minerals/That weakens motion" ("motion" here refers to perception).*

Brabantio: O thou foul thief, where hast thou stowed my daughter?
Damned as thou art, thou hast enchanted her,
For I'll refer me to all things of sense,
If she in chains of magic were not bound,
Whether a maid so tender, fair and happy,
So opposite to marriage that she shunned
The wealthy, curled darlings of our nation,
Would ever have, t'incur a general mock,
Run from her guardage to the sooty bosom
Of such a thing as thou? to fear, not to delight.
Judge me the world if 'tis not gross in sense
That thou hast practised on her with foul charms,
Abused her delicate youth with drugs or minerals
That weakens motion: I'll have't disputed on,
'Tis probable and palpable to thinking.
I therefore apprehend and do attach thee
For an abuser of the world, a practiser
Of arts inhibited and out of warrant.
Lay hold upon him; if he do resist
Subdue him at his peril!

Othello, 1.2

Othello. Jordan Baker as Desdemona and Avery Brooks as Othello.

In fact, Brabantio has no real complaint against Othello other than his race. Brabantio's racism is so simpleminded that he cannot conceive how others would not share it. Thus he acts with the confusion of a parent confronting the problems posed by a wayward child. How could his formerly dutiful daughter "Run from her guardage to the sooty bosom/Of such a thing as" Othello?

Because he cannot appreciate that Desdemona acts as she does because she perceives Othello very differently, Brabantio tries to find an external cause for her behavior. He claims that Othello has "enchanted" her and bound her in "chains of magic." But the "magic" that he fears is of a type not unknown in our time: the way in which the brain acts when "Abused . . . with drugs."

This power that drugs can have over us is just beginning to yield its secrets to brain science. Today the "chains" of drugs are felt by every addict. By definition, an addict is dependent, both emotionally and physically, on continued access to a drug, whether it is alcohol, heroin, nicotine, or another. Often, the desirable effects of the drug become progressively more difficult to achieve as the body builds a natural tolerance. But the addict must continue to use the drug to prevent the symptoms of withdrawal.

Misguided though it is, Brabantio's fear that his daughter's behavior has been altered by drugs is not difficult to understand. Love brings its own kind of dependency, although it is more emotional than physical. Trapped by her love, Desdemona cannot leave Othello, even when he becomes dangerously jealous because he mistakenly believes she has committed adultery. Among the key features of a true addiction are changes in the brain that lead to painful emotional and physical symptoms with withdrawal of the drug. In classic drug withdrawal, anxiety increases and psychotic episodes can develop, heart rate and blood pressure become unstable, and there may be profuse sweating, disturbed bowel function, and other signs of activation of the autonomic nervous system.

Identifying the changes that chronic drug use induces in the brain and the mechanisms underlying drug cravings has become

Judge me the world if 'tis not gross in sense
That thou hast practised on her with foul charms,
Abused her delicate youth with drugs or minerals
That weakens motion....

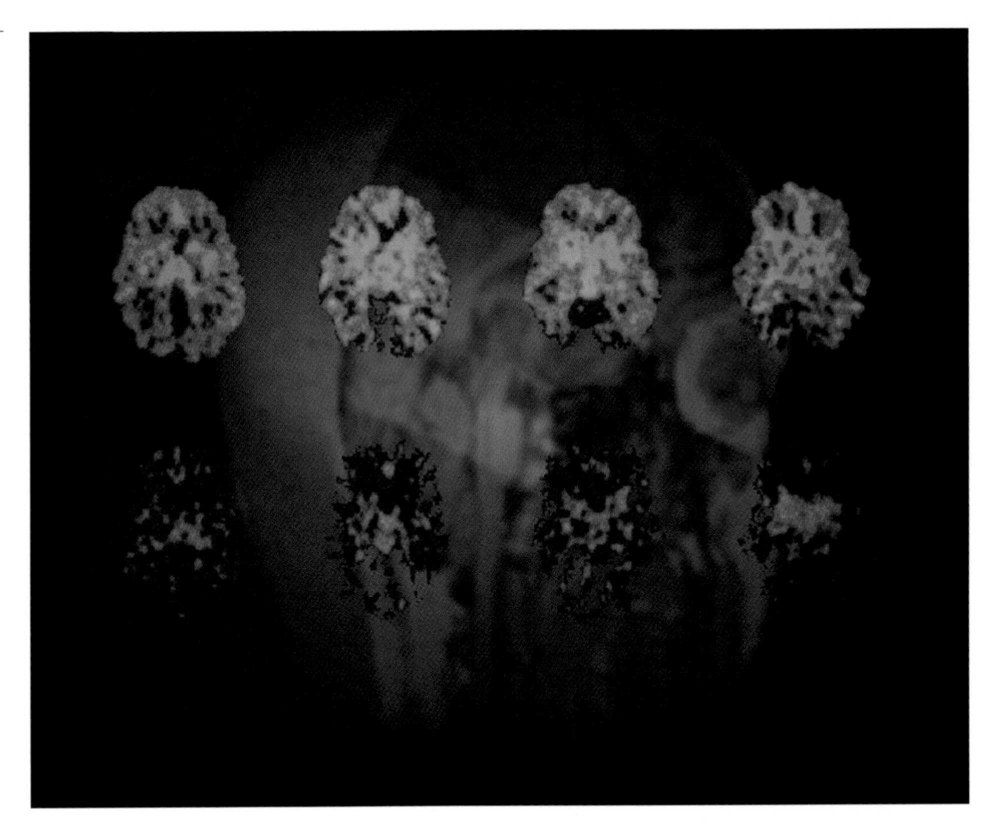

Figure 41. *Drugs used for long periods can change their own metabolism in the body and the brain. Cigarette smoking, for example, reduces the amount of a specific brain enzyme known as monoamine oxidase B. Normally, this enzyme helps to regulate neurotransmitter levels, but it also acts to break down constituents of cigarette smoke. The PET images here were generated by following the distribution of radioactively labeled selegiline (Deprenyl), a compound that binds to this enzyme. This defines areas of the brain potentially responsible for the therapeutic activity of this drug. The upper set of images are from the brain of a nonsmoker, illustrating the relatively high concentrations of monoamine oxidase B normally found within gray matter (where nerve cell bodies are found) deep in the brain. The lower images show similar views from the brain of a smoker, emphasizing the marked decrease in concentration of this enzyme that occurs with chronic tobacco use.*

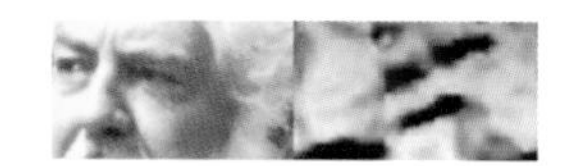

an important goal for modern brain scientists interested in understanding addiction. They are searching for the secrets of the "chains of magic."

Nicotine—found in tobacco smoke—is perhaps the most commonly abused addictive drug in the world. Understanding how nicotine alters the brain, making it so difficult for smokers to give up their habit, has been a major focus of addiction studies. An interesting example of one such study is illustrated in Figure 41. PET was used to map the distribution of an enzyme called monoamine oxidase B (MAO-B) in the brains of smokers and nonsmokers. The enzyme is a normal constituent of the brain, where it is involved in the breakdown of neurotransmitters such as dopamine and noradrenaline. The action of MAO-B therefore helps regulate neurotransmitter levels. It also acts to degrade toxic components in cigarette smoke. The much lighter color in the lower set of images (which were taken from subjects who were chronic smokers) shows that levels of this enzyme are reduced in the brains of smokers relative to those of nonsmokers. This suggests that the act of smoking itself leads to the loss of this enzyme, possibly because of a reduction in the rate of its formation in the brain. A consequence of this for chronic smokers is that quitting smoking results in a relative imbalance of brain signaling molecules. This may be a direct cause of the craving and other physical manifestations that make it so difficult to quit.

Studies of addiction may have broader relevance to understanding human behavior. Long-distance running, for example, causes the release of endorphins, the body's own opiates, which may reinforce the pleasures of exercise. The intense malaise and aching pains experienced by habitual runners forced to stop their activity may be related directly to withdrawal of this opiate stimulation. Is it too much to speculate that being loved could chronically stimulate "reward" activity in the brain, to the extent that loss of love would be perceived as a "negative reward," or punishment, causing "withdrawal" pain?

TREATING DEPRESSION

Hamlet's "To be, or not to be" soliloquy is considered by many the greatest dramatic speech ever written in English. (The fame of the speech in modern times includes its use in the Guinness Book of Records *to measure the speed of the fastest speaker, who was able to speak more than eleven words per second.) The soliloquy comes when King Claudius and Polonius have hidden nearby to eavesdrop on Hamlet's conversation with Ophelia. Before Hamlet sees her, however, he reveals his thoughts to the audience. Deeply depressed by the death of his father and the subsequent marriage of his mother to a man he despises, Hamlet grapples with the question of whether or not to continue living. However, unlike Ophelia—who eventually succumbs to depression (and probably takes her own life)—Hamlet is able to transform his mood from despondency into anger and action.*

Hamlet: To be, or not to be, that is the question:
Whether 'tis nobler in the mind to suffer
The slings and arrows of outrageous fortune,
Or to take arms against a sea of troubles
And by opposing end them. To die—to sleep,
No more; and by a sleep to say we end
The heart-ache and the thousand natural shocks
That flesh is heir to: 'tis a consummation
Devoutly to be wish'd. To die, to sleep;
To sleep, perchance to dream—ay, there's the rub:
For in that sleep of death what dreams may come,
When we have shuffled off this mortal coil,
Must give us pause—there's the respect
That makes calamity of so long life.
For who would bear the whips and scorns of time,
Th'oppressor's wrong, the proud man's contumely,
The pangs of dispriz'd love, the law's delay,

Hamlet. Sam Waterston as Hamlet.

When we have shuffled off this mortal coil, Must give us pause—there's the respect That makes calamity of so long life.

The insolence of office, and the spurns
That patient merit of th'unworthy takes,
When he himself might his quietus make
With a bare bodkin? Who would fardels bear,
To grunt and sweat under a weary life,
But that the dread of something after death,
The undiscover'd country, from whose bourn
No traveller returns, puzzles the will,
And makes us rather bear those ills we have
Than fly to others that we know not of?
Thus conscience does make cowards of us all,
And thus the native hue of resolution
Is sicklied o'er with the pale cast of thought,
And enterprises of great pitch and moment
With this regard their currents turn awry
And lose the name of action. Soft you now,
The fair Ophelia! Nymph, in thy orisons
Be all my sins remember'd.

Hamlet, 3.1

In all of world literature there is perhaps no more famous line than "To be, or not to be, that is the question." Hamlet is considering whether he should take his own life to escape his deep unhappiness. His depression has been triggered by two losses in quick succession—his father's death and his mother's hasty remarriage—for he sees his mother's remarriage as the loss of her to his uncle. Major life events such as these are common triggers of depression.

This passage is of extraordinary interest for several reasons. There is the dark beauty of Hamlet's words, which turn his personal struggle into a universal one. The importance of this soliloquy to the plot is that it marks a key moment of decision for Hamlet. Although he begins with thoughts of suicide, these "currents turn awry/And lose the name of action." With this change of heart comes a renewed desire to confront his uncle's crime directly and seek redress.

The soliloquy also provides a remarkable example of a form of personal psychological therapy. We see Hamlet rationally examining the possibility of suicide and its potential consequences, weighing the uncertainties of death against the continued struggle of life. He questions how someone as tortured as he could possibly avoid being tempted by suicide "When he himself might his quietus make/With a bare bodkin?" (*bodkin* is another word for "dagger"). But in comparing death to sleep, Hamlet becomes concerned because "in that sleep of death what dreams may come,/When we have shuffled off this mortal coil,/Must give us pause." Fear concerning the afterlife—the "dread of something after death"—stays his hand.

Medical practice today still affords psychological or behavioral therapies a role in treatment of mild or moderate depression. Recognizing depression, understanding its roots, and finding coping strategies to minimize its impact can help relieve less severe episodes. One of the extraordinary features of the mind is that it can be modified by its own actions, and by behavior and environment, as well as by drugs. In fact, psychological or behavioral therapies can provide benefits similar to those of antidepressant medications. However, for more severe depression, treatments other than those that are simply psychologically based are necessary. Drugs are the mainstay of the modern treatment of more severe depression.

While the underlying cause of depression remains unknown, it is becoming clear that people who suffer depression may have a lifelong tendency toward the illness, suggesting a difference in the way their brains work. Communication between cells in the brain occurs by means of neurotransmitter signaling molecules. Among them is the neurotransmitter serotonin. Many studies have suggested that responses to endogenous serotonin levels are reduced in the brains of patients with depression or a tendency to depression. These studies have provided impetus for the development of drugs that increase serotonin levels. The most recent such drugs work by inhibiting the cellular mechanisms that rapidly remove serotonin once it is released. Prozac

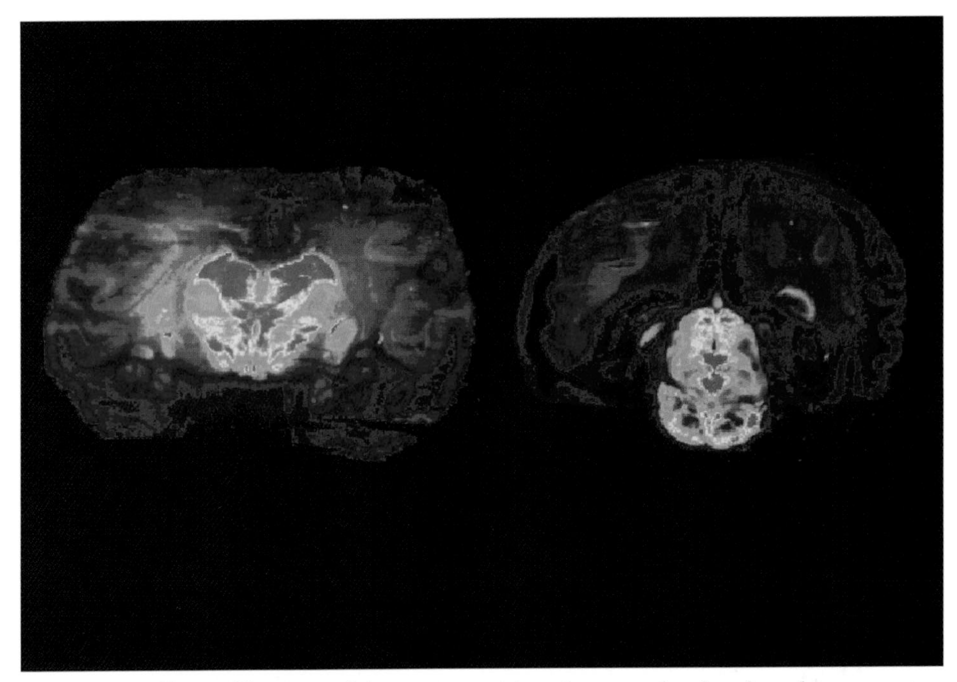

Figure 42. *One of the most exciting drugs to be developed in recent years is fluoxetine (Prozac). This powerful antidepressant is effective and has a relatively low incidence of side effects. It is thought to interact with the molecules on the outside of neurons that take up the neurotransmitter serotonin. The single-photon-emission computed tomography (SPECT) images of a monkey brain shown here were produced by means of a special radio-labeled tracer molecule that has a structure similar to that of fluoxetine. These images indicate the areas of the brain where this drug might work. On the left side is an image acquired soon after the injection of the tracer alone. On the right side is an image made after injection of both the tracer and a compound that blocks the neuronal surface receptor molecules.*

(fluoxetine) is the most widely known of this family of drugs, which also includes Zoloft (sertraline) and Paxil (paroxetine). The importance of serotonin to brain function can be appreciated in Figure 42, which shows the widespread distribution of the nerve cell surface molecule (or receptor) responsible for the uptake of serotonin into cells of the monkey brain (which is similar to the human brain in this respect). With improved techniques, the same sort of noninvasive mapping of these receptors should become possible in the living human brain.

It is possible that both psychological and behavioral (as well as drug) therapies can change the levels of molecules that control the levels of neurotransmitters such as serotonin. This would account for the fact that brief treatment regimes sometimes have long-lasting beneficial effects. Consider, for example, that despite evidence of an underlying, chronic tendency for depression in the brain of a sufferer, treatment with an antidepressant drug can often be discontinued after only months of use. While Hamlet's rational consideration of the risks and benefits of suicide as a solution to his depression is really too brief to be recognized as therapy in the true sense, it illustrates the way in which cognitive processes also can change an individual's mood.

In *Hamlet*, Shakespeare provides us with the contrast of Hamlet and Ophelia, both of whom experience depression triggered by the real or imagined loss of a loved one. Although the two cases have very different outcomes, the dark alternative of suicide is raised for both characters. Hearing his famous soliloquy, we are witnesses to Hamlet's rejection of this temptation. Ophelia's later death, however, is a tragic reminder that not all people who suffer from depression may be able to overcome their illness. It emphasizes the urgent need to improve treatments for this common disorder of the brain.

THE PROMISE OF TREATMENT

Brain imaging methods will contribute to the future development of more effective treatments for diseases of the brain and mind. In Macbeth, *the doctor reports on the worsening condition of Lady Macbeth to her distraught husband. While Macbeth himself can do nothing, his hope that medicine may be able to help is so great that even the report of approaching enemy soldiers does not distract him from his consultation with the doctor. He asks the doctor if medicine (an "antidote") might be given or whether sorrow cannot be somehow physically removed from the human memory. The doctor notes that effective treatment of a mental illness usually demands not only medical skill but also the active participation of the patient, who must "minister to himself."*

Macbeth: How does your patient, Doctor?

Doctor: Not so sick, my Lord,
As she is troubled with thick-coming fancies,
That keep her from her rest.

Macbeth: Cure her of that:
Canst thou not minister to a mind diseas'd,
Pluck from the memory a rooted sorrow,
Raze out the written troubles of the brain,
And with some sweet oblivious antidote
Cleanse the stuff'd bosom of that perilous stuff
Which weighs upon the heart?

Macbeth, 5.3

It is likely that guilt, stress, and the consequent chronic lack of sleep have triggered psychotic illness in Lady Macbeth. Manifestation of an underlying tendency to schizophrenia or

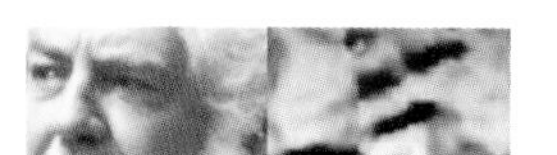

Macbeth. Helen Carey as Lady Macbeth.

Canst thou not minister to a mind diseas'd, Pluck from the memory a rooted sorrow, Raze out the written troubles of the brain, And with some sweet oblivious antidote Cleanse the stuff'd bosom of that perilous stuff Which weighs upon the heart?

depression often may be initiated by severe stresses. The heartache of Macbeth as he sees his beloved wife wandering in the grip of madness remains an all-too-common experience. Neurological and psychiatric diseases are major causes of disability and death among people of all ages, and their impact is growing as people live longer.

The sheer enormity of the problem means that much remains to be done before doctors will be able to provide a "sweet oblivious antidote" to "Cleanse the stuff'd bosom of that perilous stuff/Which weighs upon the heart." Nonetheless, the situation has changed dramatically since Shakespeare's time.

Shakespeare lived in a period in which the organic basis of neurological and psychiatric disease was only beginning to be appreciated. Thomas Willis, arguably the first doctor to view patients as a modern neurologist, would not be born until five years after Shakespeare's death. It is thus with a wonderful note of optimism that we can bring this book to a close by recalling a few of the tremendous advances that have been made in treatment as a result of improved understanding of the mechanisms of the normal and the diseased brain. Brain imaging techniques have contributed greatly to this growth in understanding and are expected to contribute even more in the future.

Before the end of the last century, patients with epilepsy, such as Julius Caesar, were left effectively untreated with recurrent fits. The first generation of anticonvulsant drugs had substantial side effects and often did not work completely. These were followed by the development of surgical approaches that could cure certain forms of epilepsy. Happily, over recent decades, many new medications have been developed that have much reduced side effects and a range of mechanisms of action that allow them to be used together to provide effective control for the majority of patients.

Our understanding of the causes of the problems in initiating and controlling movements that characterize disorders such as Parkinson's disease (which leads to uncontrollable tremor and immobility), Tourette's syndrome (which is associated

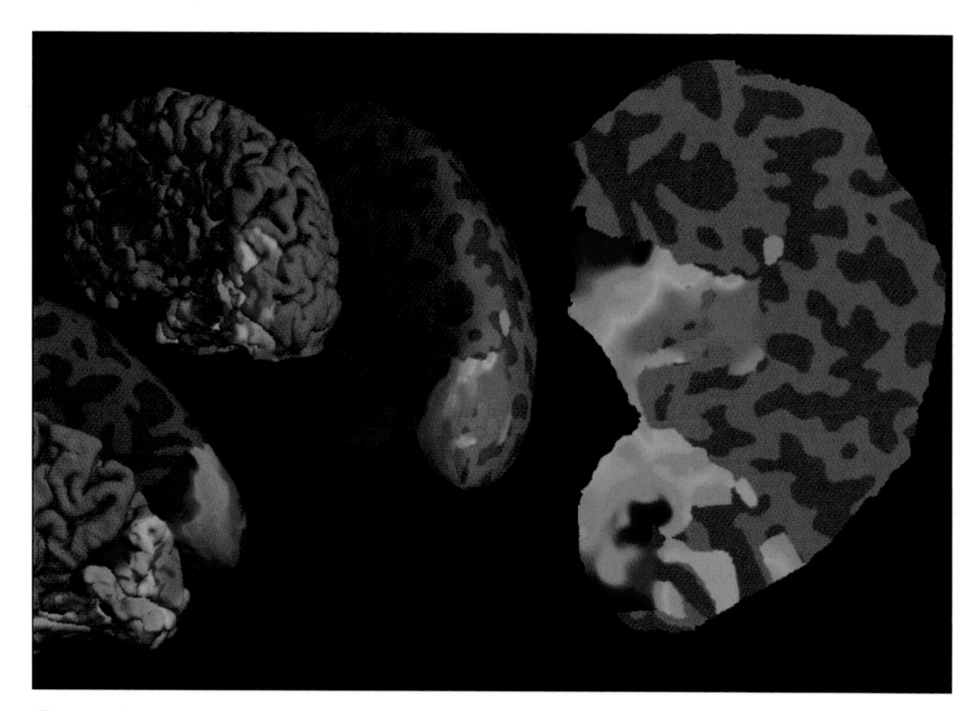

Figure 43. *Migraines are among the most common of all neurological problems. While they almost always have a benign prognosis, they are the cause of many lost hours of work and pleasure. In consequence, they have been the target of major (and successful) drug development efforts. The evolution of a migraine is uniquely illustrated in these images, which show the progression of blood flow changes in the visual cortex during the visual "aura" preceding a headache. The images on the left show a surface-rendered brain, "inflated" using computer techniques. On the right-hand side is a representation in which this brain surface has been "unfolded" to generate a so-called flat map of the data.*

with florid tics of various types), and other so-called movement disorders remains incomplete. Nonetheless, the discovery that with Parkinson's disease the cells in the brain that provide the neurotransmitter dopamine die has led to the introduction of dopamine's chemical precursor, L-dopa, as a treatment. This drug has dramatically changed the lives of sufferers because the effects of the progressive disease can be delayed for years by taking a simple pill. More recently, promising attempts have been made to transplant cells directly into the brain to replace those that have died. If this strategy can be made to work well, the disease could effectively be "cured."

The ravages of an inflammatory disorder of the brain such as multiple sclerosis (or MS, which strikes with particular tragedy in younger people) may finally slow with the introduction of a range of novel biological compounds that limit the body's ability to mount an attack on its own brain cells. Now there is real hope that with early diagnosis and treatment, the rate of progression and ultimate consequences of the disease may be modified. Similarly, with stroke, new "clot busters" have been introduced that can dramatically improve the prognosis if given soon enough after the blockage of a blood vessel.

Fortunately, not all neurological or psychiatric disease is devastating, but even less severe disorders can be troublesome. For example, one of the most common brain disorders is the migraine headache. Perhaps as many as two thirds of people suffer debilitating migraine headaches at some point in their lives, and migraine is responsible for considerable loss of productivity and income. One of the most prominent symptoms associated with the headache is visual disturbance. There can be a brief loss of even a large area of normal vision before the headache starts. The progression of the changes in brain activity associated with this have been mapped in the accompanying images made by Roger Tootell's group at Massachusetts General Hospital (Fig. 43). They found a man who reliably developed migraines every time he exercised. They asked him to begin basketball practice, and then, when he felt the first warnings of an impending headache, they began to image his brain using fMRI. As he developed a visual disturbance (partial loss of vision—a black "hole," known as a scotoma—in his visual field), they were able to map the course of changes in the visual centers of the brain. They found a clear sequence of changes in blood flow in the back part of the brain responsible for vision that paralleled the symptoms the patient experienced. The new understanding of these extraordinary phenomena, developed just in the last two to three decades, has led to exciting advances in the treatment of migraine.

Although it remains inadequate, our understanding of the molecular and cellular basis of disease (such as why brain cells die with stroke, or the special mechanisms that allow cancers

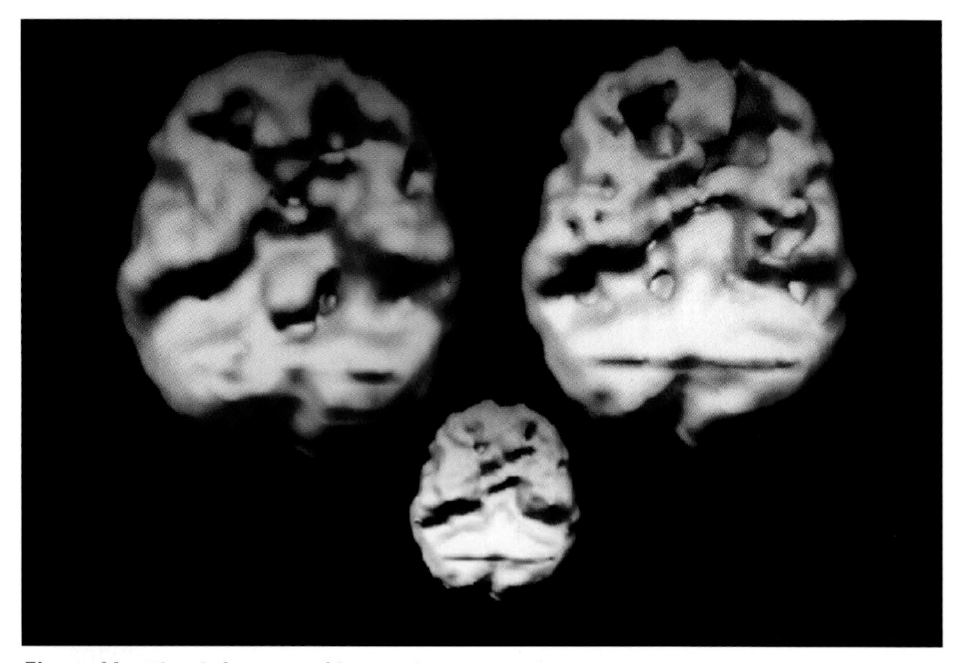

Figure 44. *The definition of brain disease is changing. Attention Deficit Disorder, for example, was previously a rare diagnosis, but it has been much more commonly recognized in recent years. One of the most important reasons for establishing a specific diagnosis is to allow the initiation of therapy to relieve symptoms. The unusual images shown here are representations of results from single-photon-emission computed tomography (SPECT) scans, which measure the relative rates of metabolism in different parts of the brain. The upper pair of images were acquired from patients with attention deficit disorder. They show mild (left) and more marked (right) decreases in metabolism in the front of the brain. Following treatment with the drug Ritalin, an amphetamine-like agent, there is an improvement in behavior, as well as the development of a more normal pattern of activity in the front of the brain.*

to maintain a good blood supply to support their rapid growth) has suggested whole new approaches to drug treatment for disorders of the brain. Even the challenge of treating psychiatric diseases is finally beginning to be met. The fundamental basis of such common diseases as depression and schizophrenia remains unknown, but empirical studies have established a large range of medications that can be used to control symptoms. In a similar way, the symptoms of attention deficit disorder, a common hyperactivity syndrome affecting children, can be controlled with the drug Ritalin. The effects of Ritalin (methylphenidate) on patterns of brain activity is reflected in SPECT scans in the brain illustration in Figure 44.

Shakespeare was neither a psychologist nor a brain scientist in any way that would fit the modern meanings of these terms, yet he was deeply interested in, and keenly observant of, the behavior of his fellow man, whether healthy or diseased. His interest led to his own literary "dissection" of the mechanisms of the mind. These insights into the fundamental issues of the mind contributed to his stature as a successful playwright in his own time and make his work a joy to read and see produced again and again even in ours, 400 years later. He clearly could not have foreseen the advances in brain science that have occurred since he lived, but surely Shakespeare's fascination with the human mind would have been further fueled by the discoveries of modern brain science.

Scientific Sources

Introduction

Posner MI, Raichle ME. The neuroimaging of human brain function. Proc Natl Acad Sci U S A. 1998;95:763–4.

Raichle ME. Visualizing the mind. Sci Am. 1994;270:58–64.

Chapter 1: Minds and Brains

Johnston MV, Nishimura A, Harum K, Pekar J, Blue ME. Sculpting the developing brain. Adv Pediatr. 2001;48:1–38.

Mega MS, Thompson PM, Cummings JL, et al. Sulcal variability in the Alzheimer's brain: correlations with cognition. Neurology. 1998;50:145–51.

Paus T, Zijdenbos A, Worsley K, et al. Structural maturation of neural pathways in children and adolescents: in vivo study. Science. 1999;283:1908–11.

Rehm K, Lakshminaryan K, Frutiger S, et al. A symbolic environment for visualizing activated foci in functional neuroimaging datasets. Med Image Anal. 1998;2:215–26.

Thompson PM, Moussai J, Zohoori S, et al. Cortical variability and asymmetry in normal aging and Alzheimer's disease. Cereb Cortex. 1998;8:492–509.

Chapter 2: Seeing, Smelling, Feeling

Gauthier I, Behrmann M, Tarr MJ. Can face recognition really be dissociated from object recognition? J Cogn. Neurosci. 1999;11:349–70.

Kanwisher N, Stanley D, Harris A. The fusiform face area is selective for faces not animals. Neuroreport. 1999;10:183–7.

Ploghaus A, Tracey I, Gati JS, et al. Dissociating pain from its anticipation in the human brain. Science. 1999;284:1979–81.

Royet JP, Koenig O, Gregoire MC, et al. Functional anatomy of perceptual and semantic processing for odors. J Cogn. Neurosci. 1999;11:94–109.

Tracey I, Becerra L, Chang I, et al. Noxious hot and cold stimulation produce common patterns of brain activation in humans: a functional magnetic resonance imaging study. Neurosci Lett 2000. Jul. 14;288(2):159–62.

Weismann M, Yousry I, Heuberger E, et al. Functional magnetic resonance imaging of human olfaction. Neuroimaging. Clin N Am 2001. May;11(2):237–50, viii.

Chapter 3: Decision and Action

Damasio H, Grabowski T, Frank R, Galaburda AM, Damasio AR. The return of Phineas Gage: clues about the brain from the skull of a famous patient. Science. 1994;264:1102–5.

Miall RC, Reckess GZ, Imamizu H. The cerebellum coordinates eye and hand tracking movements. Nat Neurosci 2001. Jun:4(6): 638–44.

Neele SJ, Rombouts SA, Bierlaagh MA, Barkhof F, Scheltens P, Netelenbos JC. Raloxifene affects brain activation patterns in postmenopausal women during visual encoding. J Clin Endocrinol Metab 2001. Mar;86(3):1422–4. 86.

Thomsen T, Hugdahl K, Ersland L, et al. Functional magnetic resonance imaging (fMRI) study of sex differences in a mental rotation task. Med Sci Monit. 2000. Nov–Dec;6:1186–96.

Chapter 4: Language and Numbers

Dehaene S, Spelke E, Pinel P, Stanescu R, Tsivkin S. Sources of mathematical thinking: behavioral and brain-imaging evidence. Science. 1999;284:970–4.

Duzel E, Cabeza R, Picton TW, et al. Task-related and item-related brain processes of memory retrieval. Proc Natl Acad Sci U S A. 1999;96:1794–9.

Grabowski TJ, Damasio H, Damasio AR. Premotor and prefrontal correlates of category-related lexical retrieval. Neuroimage. 1998;7:232–43.

Kimchi R. The perceptual organization of visual objects: a microgenetic analysis. Vision Res 2000;40(10–12):1333–47. 40.

Stanescu-Cosson R, Pinel P, van De Moortele PF, Le Bihan D, Cohen L, Dehaene S. Understanding dissociations in dyscalculia: a brain imaging study of the impact of number size on the cerebral networks for exact and approximate calculation. Brain 2000. Nov;123(Pt. 11):2240–55. 123.

Tulving E. Episodic memory: from mind to brain. Annu Rev Psychol 2002;53:1–25.

Chapter 5: Our Inner World

Blair RJ, Morris JS, Frith CD, Perrett DI, Dolan RJ. Dissociable neural responses to facial expressions of sadness and anger. Brain. 1999;122:883–93.

Blood AJ, Zatorre RJ, Bermudez P, Evans AC. Emotional responses to pleasant and unpleasant music correlate with activity in paralimbic brain regions. Nat Neurosci. 1999;2:382–7.

Chen W, Kato T, Zhu XH, Ogawa S, Tank DW, Ugurbil K. Human primary visual cortex and lateral geniculate nucleus activation during visual imagery. Neuroreport. 1998;9:3669–74.

Giedd JN, Blumenthal J, Jeffries NO, et al. Brain development during childhood and adolescence: a longitudinal MRI study. Nat Neurosci. 1999;2:861–3.

Paus T, Zijdenbos A, Worsely K, et al. Structural maturation of neural pathways in children and adolescents: in vivo study. Science. 1999;283:1908–11.

Peretz I, Ayotte J, Zatorre RJ, et al. Congenital amusia: a disorder of fine-grained pitch discrimination. Neuron 2002. Jan. 17;33(2):185–91.

Platel H, Price C, Baron JC, et al. The structural components of music perception. A functional anatomical study. Brain. 1997;120:229–43.

Ploghaus A, Tracey I, Gati JS, et al. Dissociating pain from its anticipation in the human brain. Science. 1999;284:1979–81.

Chapter 6: The Seventh Age of Man: Disease, Aging, and Death

Bonmassar G, Schwartz DP, Liu AK, Kwong KK, Dale AM, Belliveau JW. Spatiotemporal brain imaging of visual-evoked activity using interleaved EEG and fMRI recordings. Neuroimage 2001. Jun;13(6 Pt. 1):1035–43.

Doudet DJ, Jivan S, Ruth TJ, Holden JE. Density and affinity of the dopamine D2 receptors in aged symptomatic and asymptomatic MPTP-treated monkeys: PET studies with [11C]raclopride. Synapse 2002. Jun. 1;44(3):198–202.

Liotti M, Mayberg HS. The role of functional neuroimaging in the neuropsychology of depression. J Clin Exp Neuropsychol 2001. Feb;23(1):121–36.

Mayberg HS, Liotti M, Brannen SK, et al. Reciprocal limbic-cortical function and negative mood: converging PET findings in depression and normal sadness. Am J Psychiatry. 1999;156:675–82.

Narayanan S, Fu L, Pioro E, et al. Imaging of axonal damage in multiple sclerosis: spatial distribution of magnetic resonance imaging lesions. Ann Neurol. 1997;41:385–91.

Shergill SS, Bullmore E, Simmons A, Murray R, McGuire P. Functional anatomy of auditory verbal imagery in schizophrenic patients with auditory hallucinations. Am J Psychiatry 2000. Oct;157(10):1691–3.

Staley JK, Tamagnan G, Baldwin RM, et al. SPECT imaging with the D(4) receptor antagonist L-750,667 in nonhuman primate brain. Nucl. Med Biol 2000. Aug;27(6):547–56. 27.

Stoessl AJ. Neurochemical and neuroreceptor imaging with PET in Parkinson's disease. Adv Neurol 2001; 86:215–23.

Thobois S, Guillouet S, Broussolle E. Contributions of PET and SPECT to the understanding of the pathophysiology of Parkinson's disease. Neurophysiol Clin 2001. Oct;31(5):321–40.

Thompson PM, Mega WS, Woods RP, et al. Cortical change in Alzheimer's disease detected with a disease-specific population-based brain atlas. Cereb Cortex 2001. Jan;11(1):1–16.

Woodruff PW, Wright IC, Bullmore ET, et al. Auditory hallucinations and the temporal cortical response to speech in schizophrenia: a functional magnetic resonance imaging study. Am J Psychiatry. 1997;154:1676–82.

Chapter 7: Drugs and the Brain

Dager SR, Friedman SD. Brain imaging and the effects of caffeine and nicotine. Ann Med 2000. Dec;32(9):592–9.

Due DL, Huettel SA, Hall WG, Rubin DC. Activation in mesolimbic and visuospatial neural circuits elicited by smoking cues: evidence from functional magnetic resonance imaging. Am J Psychiatry 2002. Jun;159(6):954–60.

Hadjikhani N, Sanchez DR, Wu O, et al. Mechanisms of migraine aura revealed by functional MRI in human visual cortex. Proc Natl Acad Sci U S A 2001. Apr. 10;98(8):4687–92.

Kahkonen S, Kesaniemi M, Nikouline VV, et al. Ethanol modulates cortical activity: direct evidence with combined TMS and EEG. Neuroimage 2001. Aug;14(2):322–8.

Seifritz E, Bilecen D, Hanggi D, et al. Effect of ethanol on BOLD response to acoustic stimulation: implications for neuropharmacological fMRI. Psychiatry Res 2000. Jul 10;99(1):1–13.

Credits

FIGURES

*Page 17, Fig 1:*Y.G. Abdullaev, and M.I. Posner, Neuroimage 7 (1998) 1–13.

Page 18, Fig 2: Courtesy: Keith Worsley, Department of Mathematics and Statistics, McGill University

Page 19, Fig 3: Heidi Johansen-Berg FMRIB Centre, University of Oxford

Page 21, Illustration copyright Kathryn Born

Page 25, Fig 4: Courtesy: Kelly Rehm, Department of Radiology, University of Minnesota

Page 32, Fig 5: Courtesy: Tom Bowles, Department of Radiology, University of Oxford

Page 33, Fig 6: Courtesy: Paul Thompson, UCLA School of Medicine

Page 38, Fig 7: Courtesy: Alan Evans, Montreal Neurological Institute, Montreal Neurological Institute, McConnell Brain Imaging Center

Page 38, Mauritshuis Museum, The Hague, Netherlands (background Rembrandt)

Page 39, Fig 8: Courtesy: Roberto Pineiro, FMRIB Centre, University of Oxford

Page 45, Fig 9: Courtesy: Eric Halgren, Massachusetts General Hospital, Harvard Medical School

Page 51, Fig 10: Courtesy: Robert Osterbauer, FMRIB Centre, University of Oxford

Page 57, Fig 11: Courtesy: Irene Tracey, FMRIB Centre, Oxford University

Page 63, Fig 12: Courtesy: Kenneth Hugdahl, Department of Biological and Medical Psychology, University of Bergen

Page 69, Fig 13: Courtesy: Chris Miall, University Laboratory of Physiology, Oxford University

Page 74, Fig 14: Courtesy: Chris Miall, University Laboratory of Physiology, Oxford University

Page 80, Fig 15: Courtesy: Adrian Owen, MRC Cognition and Brain Sciences Unit, Cambridge

Page 86, Fig 16: H. Damasio et al *Science* 264 (1994): 1102–5

Page 92, Fig 17: SW Anderson, et al. *Nature Neuroscience* 2 (1999); 1032-7

Page 93, Fig 18: Courtesy Adrian Raine, University of Southern California

Page 98, Fig 19: E. Halgren et al, *NeuroImage* 7 (1998) 232–43

Page 100, Fig 20: Courtesy: Marcus Raichle, Washington University

Page 104, Fig 21: E. Halgren et al. *NeuroImage* 7 (1998): 232–43

Page 109, Fig 22: R. Cabeza R et al. *Journal of Cognitive Neuroscience* 9 (1997): 254–65

Page 115, Fig 23: T.J. Grabowski et al. *NeuroImage* 7 (3) (April 1998): 234–43

Page 120, Fig 24: K. Kim et al. *Nature* 388 (1997): 171–73

Page 125, Fig 25: S. Dehaene, S et al. *Science* 7(284) (May 1999): 970-84

Page 134, Fig 26: Courtesy: A. Castellano-Smith, Guy's Hospital Medical School

Page 135, Fig 27: Courtesy: Paul Thompson, UCLA School of Medicine

Page 140, Fig 28: Courtesy: RSJ Frackowiak, Institute of Neurology UCL

Page 145, Fig 29: A.J. Blood, et al. *Nature Neuroscience* 2 (1999): 383–87

Page 150, Fig 30: JS Morris et al, *Nature,* 383, (1996): 812–15

Page 155, Fig 31: Irene Tracey, FMRIB Centre, Unversity of Oxford

Page 160, Fig 32: Courtesy: Kamil Ugurbil, CMRR, University of Minnesota

Page 167, Fig 33: Courtesy Paul Thompson, UCLA School of Medicine, and FMRIB Center, University of Oxford

Page 173, Fig 34: Courtesy: Laboratoire IDM, Université de Rennes 1. France, http://idm.univ-rennes1.fr

Page 174, Fig 35: Courtesy: Laboratoire IDM, Université de Rennes 1. France, http://idm.univ-rennes1.fr

Page 179, Fig 36: Courtesy SHFJ/CEA/Jean-François Mangin

Page 186, Fig 37: H. Mayberg, *Am. J. Psych.* 156 (1999): 675–82.

Page 191, Fig 38: Courtesy: Katherine Narr and Arthur W Toga, Laboratory of Neuro Imaging, UCLA School of Medicine

Page 197, Fig 39: Courtesy: Stuart Clare, FMRIB Centre, University of Oxford

Page 203, Fig 40: Courtesy: Irene Tracey, FMRIB Centre, University of Oxford

Page 208, Fig 41: Joanna S Fowler, Brookhaven National Laboratory

Page 214, Fig 42: Anat Biegon. Center for Functional Imaging, Lawrence Berkeley Laboratory

Page 219, Fig 43: Courtesy N Hordjikhani, MGH NMR Centre, Harvard

Page 221, Fig 44: Courtesy: Daniel G. Amen, Amen Clinics

PERFORMANCE IMAGES

Page 5: ***Romeo and Juliet***
The Shakespeare Theatre, Washington, DC, 1986; Derek Smith (Romeo), Laura Hicks (Juliet), directed by Michael Kahn; photograph by Joan Marcus.

Page 6: ***Romeo and Juliet***
The Shakespeare Theatre, Washington, DC, 1986; Derek Smith (Romeo), Edward Gero (Tybalt), directed by Michael Kahn; photograph by Joan Marcus.

Page 10: ***Hamlet***
Martha Swope/TimePix, 1990; Kevin Kline (Hamlet)

Page 24: ***The Tempest***
The Shakespeare Theatre, Washington, DC, 1997; Ana Reeder (Miranda), Ted van Griethuysen (Prospero), directed by Garland Wright; photograph by Carol Rosegg.

Page 30: ***Hamlet***
Martha Swope/TimePix, 1986; Kevin Kline (Hamlet).

Page 36: ***The Merchant of Venice***
The Phoenix Theatre, London, 1989; Geraldine James (Portia), Dustin Hoffman (Shylock); Copyright Donald Cooper/Photostage.

Page 42: ***Romeo and Juliet***
Stratford-upon-Avon, 2000; Alexandra Gilbreath (Juliet), David Tennant (Romeo); Copyright Donald Cooper/Photostage.

Page 48: ***Hamlet***
Copyright George E. Joseph; Sam Waterston (Hamlet), Charles Cioffi (Claudius), 1976.

Page 54: ***Hamlet***
The Shakespeare Theatre, Washington, DC, 1992; Tom Hulce (Hamlet), Paul Mullins (Rosencrantz), J.C. Cutler (Guildenstern), directed by Michael Kahn; photographs by Joan Marcus.

Page 61: ***The Taming of the Shrew***
Martha Swope/TimePix, 1990, Tracey Ullman (Katherina), Morgan Freeman (Petruchio).

Page 66: ***Henry V***
The Shakespeare Theatre, Washington, DC, 1995; Harry Hamlin (Henry V), directed by Michael Kahn; photograph by Carol Rosegg.

Page 72: ***Macbeth***
The Shakespeare Theatre, Washington, DC, 1995; Stacy Keach (Macbeth), directed by Joe Dowling, photograph by Carol Rosegg.

Page 77: ***Love's Labour's Lost***
The Shakespeare Theatre, Washington, DC, 1995; Clockwise from upper left: Jason Patrick Bowcutt, Alene Dawson, Libby Christophersen, Michael Medico, Enid Graham, Dallas Roberts, Melissa Bowen and Sean Pratt, directed by Laird Williamson; photograph by Carol Rosegg.

Page 84: ***Henry IV, Part I***
The Shakespeare Theatre, Washington, DC, 1994; Derek Smith (Prince Hal), directed by Michael Kahn; photograph by Carol Pratt.

Page 89: ***Richard III***
The Shakespeare Theatre, Washington, DC, 1990; Stacy Keach (Richard III), directed by Michael Kahn; photograph by Joan Marcus.

Page 97: ***Henry V***
The Shakespeare Theatre, Washington, DC, 1995; Wallace Acton (Chorus member), directed by

Michael Kahn; photograph by Carol Pratt.

Page 102: ***The Merchant of Venice***
The Shakespeare Theatre, Washington, DC, 1999; Enid Graham (Portia), Hal Holbrook (Shylock), directed by Michael Kahn; photograph by Carol Rosegg.

Page 107: ***Henry IV, Part I***
Copyright George E. Joseph, 1968; Sam Waterston (Henry V), Stacy Keach (Falstaff).

Page 112: ***Romeo and Juliet***
The Shakespeare Theatre, Washington, DC, 1986; Laura Hicks (Juliet), directed by Michael Kahn; photograph by Joan Marcus.

Page 119: ***Henry V***
The Shakespeare Theatre, Washington, DC, 1995; Vivienne Benesch (Katherine), Harry Hamlin (Henry V), directed by Michael Kahn; photograph by Carol Pratt.

Page 123: ***Julius Caesar***
Copyright George E. Joseph, 1988; Martin Sheen (Brutus).

Page 130: ***The Tempest***
The Shakespeare Theatre, Washington, DC, 1997; Ana Reeder (Miranda), Ted van Griethuysen (Prospero), directed by Garland Wright; photograph by Carol Rosegg.

Page 138: ***The Winter's Tale***
Martha Swope/TimePix, 1989; (L-R) Mandy Patinkin, Alfre Woodard, James Olson, Diane Venora, Christopher Reeve, Jennifer Dundas, and Graham Winton.

Page 143: ***Twelfth Night***
The Shakespeare Theatre, Washington, DC, 1998; Michael Rudko (Feste), Graham Winton (Orsino), directed by Daniel Fish; photograph by Carol Rosegg.

Page 148: ***King Lear***
The Shakespeare Theatre, Washington, DC, 1999; Monique Holt (Cordelia), Ted van Griethuysen (Lear), directed by Michael Kahn; photograph by Carol Rosegg.

Page 153: ***Macbeth***
The Shakespeare Theatre, Washington, DC, 1995; Stacy Keach (Macbeth), Chris McKinney (Macduff), directed by Joe Dowling; photograph by Carol Pratt.

Page 158: ***The Tempest***
West Yorkshire Playhouse, Leeds, England, 1999; Claudie Blakely (Miranda), Ian McKellen (Prospero), Rashan Stone (Ferdinand); Copyright Donald Cooper/Photostage.

Page 164: ***As You Like It***
The Shakespeare Theatre, Washington, DC, 1997; Floyd King (Jaques), directed by Laurence Boswell; photograph by Carol Rosegg.

Page 171: ***Julius Caesar***
Stratford-upon-Avon, 1995; Copyright Donald Cooper/Photostage; John Nettles (Brutus), Julian Glover (Cassius).

Page 177: ***Henry VIII***
Martha Swope/TimePix, 1976; Nicol Williamson (Henry VIII).

Page 183: ***Richard II***
Shakespeare Centre Library, Stratford-Upon-Avon; Jeremy Irons (Richard II), 1986.

Page 189: ***King Lear***
Olivier Theatre/National Theatre, London, 1986; Anthony Hopkins (Lear); Copyright Donald Cooper/Photostage.

Page 195: ***Henry IV, Part II***
The Shakespeare Theatre, Washington, DC, 1994; David Sabin (Falstaff), Derek Smith (Hal), directed by Michael Kahn; photograph by Carol Pratt.

Page 200: ***Romeo and Juliet***
The Shakespeare Theatre, Washington, DC, 1986; Emery Battis (Capulet), Fran Dorn (Lady Capulet), Laura Hicks (Juliet), directed by Michael Kahn; photograph by Joan Marcus.

Page 206: ***Othello***
The Shakespeare Theatre, Washington, DC, 1991; Jordan Baker (Desdemona), Avery Brooks (Othello), directed by Harold Scott; photograph by Joan Marcus.

Page 211: ***Hamlet***
Copyright 1976, George E. Joseph; Sam Waterston (Hamlet).

Page 217: ***Macbeth***
The Shakespeare Theatre, Washington, DC, 1995; Helen Carey (Lady Macbeth), directed by Joe Dowling; photograph by Carol Rosegg.

Index